The Sky's the Limit!

AUTHORS

Stan Adair
Dennis Ivans
Bill Shennan

Bill Barker
Zachary Sconiers
Marney Welmers

Dave Youngs

EDITORS

Arthur Wiebe

Larry Ecklund

Betty Cordel

ILLUSTRATOR

Sheryl Mercier

Margo Pocock

TECHNICAL ILLUSTRATOR

Johann Weber

This book contains materials developed by the AIMS Education Foundation. **AIMS** (Activities Integrating Mathematics and Science) began in 1981 with a grant from the National Science Foundation. The non-profit AIMS Education Foundation publishes hands-on instructional materials (books and the monthly *AIMS* Magazine) that integrate curricular disciplines such as mathematics, science, language arts, and social studies. The Foundation sponsors a national program of professional development through which educators may gain both an understanding of the AIMS philosophy and expertise in teaching by integrated, hands-on methods.

ISBN 1-881431-44-4

Printed in the United States of America

Project 2061 Benchmarks

- Results of similar scientific investigations seldom turn out exactly the same. Sometimes this is because of unexpected differences in the things being investigated, sometimes because of unrealized differences in the methods used or in the circumstances in which the investigation is carried out, and sometimes just because of uncertainties in observations. It is not always easy to tell which.

- When similar investigations give different results, the scientific challenge is to judge whether the differences are trivial or significant, and it often takes further studies to decide. Even with similar results, scientists may wait until an investigation has been repeated many times before accepting the results as correct.

- Results of scientific investigations are seldom exactly the same, but if the differences are large, it is important to try to figure out why. One reason for following directions carefully and for keeping records of one's work is to provide information on which might have caused the differences.

- If more than one variable changes at the same time in an experiment, the outcome of the experiment may not be clearly attributable to any one of the variables. It may not always be possible to prevent outside variables from influencing the outcome of an investigation (or even to identify all of the variables), but collaboration among investigators can often lead to research designs that are able to deal with such situations.

- Clear communication is an essential part of doing science. It enables scientists to inform others about their work, expose their ideas to criticism by other scientists, and stay informed about scientific discoveries around the world.

- Doing science involves many different kinds of work and engages men and women of all ages and backgrounds.

- Mathematics provides a precise language for science and technology–to describe objects and events, to characterize relationships between variables, and to argue logically.

- Measuring instruments can be used to gather accurate information for making scientific comparisons of objects and events and for designing and constructing things that will work properly.

- There is no perfect design. Designs that are best in one respect (safety or ease of use, for example) may be inferior in other ways (cost or appearance). Usually some features must be sacrificed to get others. How such trade-offs are received depends upon which features are emphasized and which are down-played.

- Even a good design may fail. Sometimes steps can be taken ahead of time to reduce the likelihood of failure, but it cannot be entirely eliminated.

- The solution to one problem may create other problems.

- The earth's gravity pulls any object toward it without touching it.

- Learning means using what one already knows to make sense out of new experiences or information, not just storing the new information in one's head.

- *When people care about what is being counted or measured, it is important for them to say what the units are (three degrees Fahrenheit is different from three centimeters, three miles from three miles per hour).*

- *Measurements are always likely to give slightly different numbers, even if what is being measured stays the same.*

- *Tables and graphs can show how values of one quantity are related to values of another.*

- *Many objects can be described in terms of simple plane figures and solids. Shapes can be compared in terms of concepts such as parallel and perpendicular, congruence and similarity, and symmetry. Symmetry can be found by reflection, turns, or slides.*

- *The scale chosen for a graph or drawing makes a big difference in how useful it is.*

Students should:
- *Keep records of their investigations and observations and not change the records later.*

- *Offer reasons for their findings and consider reasons suggested by others.*

- *Know why it is important in science to keep honest, clear, and accurate records.*

- *Know that often different explanations can be given for the same evidence, and it is not always possible to tell which one is correct.*

- *Add, subtract, multiply, and divide whole numbers mentally, on paper, and with a calculator.*

- *Use fractions and decimals, translating when necessary between decimals and commonly encountered fractions.*

- *Judge whether measurements and computations of quantities such as length, area, volume, weight, or time are reasonable in a familiar context by comparing them to typical values.*

- *Find what percentage one number is of another and figure any percentage of any number.*

- *Use, interpret, and compare numbers in several equivalent forms such as integers, fractions, decimals, and percents.*

- *Find the mean and median of a set of data.*

- *Recognize when comparisons might not be fair because some conditions are not kept the same.*

- *Be aware that there may be more than one good way to interpret a given set of findings.*

	Observing	Predicting	Collecting/ Recording Data	Interpreting Data	Applying/ Generalizing	Controlling Variables	Geometry	Measurement	Estimation	Averaging	Using Formulas
Seeing is Believing	x	x			x						
How High Is It?	x		x	x				x	x		
How High Can You Throw?	x		x	x	x			x			x
Unbelievable Flying Objects	x	x			x	x	x	x			
More Unbelievable Flying Objects	x	x			x		x				
But Will It Fly?	x		x	x	x		x	x			x
Be a Rotor Promotor	x	x	x		x	x		x			x
It's a Real Corker!	x		x	x	x	x	x	x		x	x
Ah Chute!	x		x	x	x			x		x	x
Rocket Balloons #1 (Horizontal flight)	x		x	x		x		x		x	
#2 (Angled flight)	x		x	x	x	x		x		x	
#3 (Opening size variable)	x		x	x	x	x	x	x		x	
Hover Craft	x	x	x	x	x	x		x	x	x	
Water Rockets	x	x	x	x	x	x	x	x	x		x
It's the Last Straw	x	x	x	x	x	x		x		x	
This is Definitely the Last Straw	x		x	x	x	x		x		x	
Distance Data	x	x	x	x	x		x	x	x		
Length Aloft	x	x	x	x	x		x	x	x		
On Your Mark–Accuracy	x	x	x	x	x			x	x		
Train Your Plane–Aerobatics	x	x	x	x	x			x			
Super Tubes	x										
The Whole Kite and Kaboodle	x	x	x	x	x		x	x		x	x
Bernoulli Was a Bird Brain	x	x	x		x		x	x			x

Table of Contents

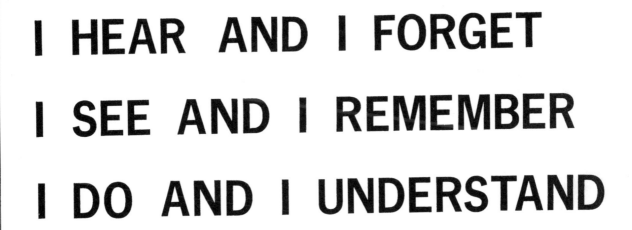

I HEAR AND I FORGET

I SEE AND I REMEMBER

I DO AND I UNDERSTAND

–Chinese Proverb

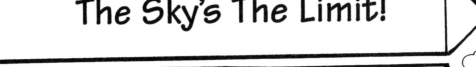

Introduction

As the title of this book implies, the sky *is* the limit, if there is one, to the knowledge and experience that can be gained by doing these investigations. As simple and inexpensive as they are, these activities explore virtually every aspect of the science of aerodynamics.

Most investigations contained in *The Sky's the Limit* were designed primarily for grades five through nine, but they can easily be adapted for use with a wider range of students.

Whether they are ready for an intuitive grasp of a general concept or the mastery of sophisticated scientific theories, students will be challenged to observe, predict, test, and generalize. Beginning with a specific question, each investigation tempt the students to open their minds, draw upon a variety of experiences, and explore a new range of possibilities, while simultaneously motivating them to support their ideas with precise data and accurate calculations.

The real beauty of these activities is that almost all provide practice and application of the basic skills in a problem-solving situation — a major thrust of education today. You probably will hear, "Gee, we didn't even have to do any math today," after your students have just completed an entire page of computing averages. Furthermore, with each activity, they will become more comfortable with the tools of science and mathematics, both the equipment and records which help us to "chart" and understand our world, which will help them to "map" and build their world.

One word of caution, however, before you take off: many students may approach you begging for extra materials from the day's investigation, so that they can set up at home and do it again! Students are indeed interested and do become involved.

Above all, *The Sky's the Limit* tries to speak to the dreamer and the explorer, the child who dares to wonder, which is in all of us, Daedalus, daVinci, Verne, and Goddard; the hang glider, the helicopter, the Concorde, and the Columbia — man's fascination with flight is eternal even though his knowledge is always increasing. Each frontier crossed offers a new frontier beyond! So it should be with this book: each question answered should generate new questions; each problem solved should create a new problem to explore. Flying is "magic" for children of all ages! This book attempts to clarify some of the reasons why the magic works while preserving its sense of mystery. As Albert Einstein expressed it: "The most beautiful thing we can experience is the mysterious. It is the source of all true art and science."[1] Flight is beauty, art, and poetry; it is science, math, and engineering. It is man's aspirations and inspirations, his past and his future, his dreams and his hope — a mystery worthy of a child's curiosity.

"No bird can soar too high if he soars with his own wings."

[1]"What I Believe," *Forum*, October, 1930.

Topic
Introduction to flight principles

Key Question
See *Background Information* for specific questions for each activity

Focus
Using a series of discrepant events, we will explore a variety of aerodynamic concepts.

Science
Project 2061 Benchmarks
- *Offer reasons for their findings and consider reasons suggested by others.*

Physical science
  aerodynamics

Processes
Observing
Predicting
Generalizing

Materials
Newspaper
Old yardstick or wood lath (available at a lumber yard)
8 1/2" x 11" paper
Various-sized balls
Thread spool
Straight pins
Candle
3" x 5" index cards

Background Information
The activities in this lesson should leave the students wondering. The outcome of each part of the experience may seem to go against common sense. These "discrepant" events are keys which will open discussions and lead to future investigation.

Newspaper Lift
Air, like all matter, has mass and exerts pressure on objects. At sea level, this pressure is roughly 15 pounds per square inch (15 psi). In this activity, students will see a dramatic demonstration of this fact. If one calculates the weight of air on a single sheet of newspaper, you can see how significant this weight can be (length 28" x width 24" x 15 psi = 10,080 lbs).
Key Question: What will happen when I hit this stick?

Paper Blow
As the student blows across the top of a piece of paper, it will be "lifted" up. This may seem strange to the students until they understand that as air moves faster across the upper surface of the paper, the pressure is reduced; the higher air pressure underneath pushes up the paper.
Key Question: What will happen when I blow across the top of this piece of paper?

Ball Drop
We have Galileo to thank for this activity. In this classic investigation, we see that gravity acts on all objects with the same force, no matter what the mass or size of the objects. The balls will strike the ground at the same time. Stress the importance of dropping the balls at precisely the same instant.
Key Question: What will happen if I drop the big ball and the little ball at the same time from the same height?

Spool and Card Blow
"Logically," if you blow air through a tube, it should push an object away from you. (Consider the lowly peashooter.) In this activity, students find that this "...ain't necessarily so." As the air spreads out between the card and spool, it moves faster than the surrounding air, thus lowering the pressure on the upper surface of the card. The students will find that the harder they blow, the more firmly the card is pushed up against the spool.
Key Question: What will happen to the card if I blow through the hole in the spool of thread?

Candle and Card Blow
This is one more example of moving air producing a low pressure area. As the student blows on the card, the air rushes outward from the center of the card. This increases the speed of the air at the edges of the card, creating a partial vacuum that pulls air from behind the candle flame. As air moves to fill the space left, the flame is bent toward the card.
Key Question: If I put this card in front of the candle's flame and blow, what will happen?

Management
1. As a starting activity, this is a teacher demonstration using student helpers.
2. This will take one class period or approximately one hour.

Procedure

1. Collect all materials listed in the materials section.
2. Ask the students to predict the outcome prior to each of the activities.
3. Solicit student helpers to do each activity in front of the class. Repeat the activity as many times as necessary for students to "see and believe!"

Discussion

1. How could one sheet of newspaper be stronger than a piece of wood?
2. What made the paper rise when we blew over it?
3. Why doesn't the heavier ball hit the ground first?
4. What held the card to the spool when we blew through the spool?
5. Why does the card get closer to the spool the harder we blow?
6. What caused the candle flame to move toward the card?
7. Can you think of any ways to vary these activities to show the same principles? What are they?

THE SKY'S THE LIMIT!

SEEING IS BELIEVING!

NAME _____

WHAT WILL HAPPEN IF.....	YOUR PREDICTION	YOUR OBSERVATION
YOU HIT THE RULER HARD? TABLE NEWSPAPER RULER		
YOU BLOW ON THE PAPER? PAPER		
YOU DROP BOTH AT THE SAME TIME FROM THE SAME HEIGHT? HARD BALL JACKS BALL		
YOU BLOW IN HERE AND LET GO OF THE CARD? SPOOL CARD PIN		
YOU BLOW HERE? WHAT WILL THE FLAME DO? CARD		

HOW HIGH IS IT?

Topic
Measuring with a clinometer

Key Question
How tall are various objects on our campus?

Focus
A clinometer will be used to find the heights of various objects.

Math
NCTM Standards
- *Generalize solutions and strategies to new problem situations*
- *Acquire confidence in using mathematics meaningfully*
- *Extend their understanding of measurement*
- *Estimate, make, and use measurements to describe and compare phenomena*

Measurement
   length
   angles
Estimating

Science
Project 2061 Benchmarks
- *Results of scientific investigations are seldom exactly the same, but if the differences are large, it is important to try to figure out why. One reason for following directions carefully and for keeping records of one's work is to provide information on which might have caused the differences.*
- *Measuring instruments can be used to gather accurate information for making scientific comparisons of objects and events and for designing and constructing things that will work properly.*
- *Measurements are always likely to give slightly different numbers, even if what is being measured stays the same.*
- *Tables and graphs can show how values of one quantity are related to values of another.*

Processes
Obeserving
Collecting and organzing data
Interpreting data

Materials
Clinometer patterns
Tagboard
Rulers
Thread
Washers or large paper clip
Strings marked in meters or meter tapes (15-30 meters long)
Tangent of an Angle (from *Water Rockets* activity)

Background Information
The height of perpendicular objects can be calculated using the laws of trigonometry. If you know the distance from your sighting point to an object and the angle from the ground to the top of the object, you calculate the height of the object by multiplying the tangent of that angle times your distance from the object. The clinometer is the tool which gives you that angle.

If the student is 15 meters from an object, the height of that object can be calculated by measuring the angle and using the *Tangent of an Angle* (from *Water Rockets* activity). If the student using the clinometer is standing up, his/her height from eye to the ground must be added to the height calculated from the tangent table.

Management
1. This activity requires 30-40 minutes, depending on the number of objects being measured.
2. Divide class into groups of four to six students.
3. Each group needs a clinometer, string or tape measure, and student recording sheet.
4. Students can rotate through the jobs of clinometer sighter, clinometer reader, data recorder and distance measurer.
5. Have students sight clinometer close to the ground by crouching or lying down; otherwise, they will have to add their heights to the calculated height of the object (see *Background Information*).
6. Have objects to be measured clearly marked beforehand. Students can rotate from one object to the next.
7. While outside, students should only record the angles of the objects and the distances from the object. They can calculate the height from the tangent table after they return to the classroom.

8. Objects such as flag poles and pointed trees are easier to sight than rounded tree tops.
9. For maximum accuracy, students must sight directly along the top of the clinometer.

Procedure

1. Assign groups to objects to be measured and explain rotation procedures.
2. Groups will proceed to their first object and estimate its height, recording their estimate on the activity sheet. Then they will use the string or tape measure to determine the distance from their sighting point to the object. They will record this distance on the table.
3. The clinometer sighter will get close to the ground at the sighting point and sight along the clinometer to the top of the object.
4. The clinometer reader will read the angle from the clinometer, and the recorder will record the angle. Then the group will move onto the next object.
5. Repeat the procedure until all the groups have collected their data.
6. Return to the classroom and calculate the heights of the objects by using the tangent table. Record the heights on the activity sheet and check to see how close the estimates were.

Discussion

1. What value do you see in this lesson? How else can we use it? [Measure altitudes of objects in the sky; measure heights of distant objects.]
2. Why did you need to lie down or kneel when you measured the angle? [If you are standing, your height must be taken into consideration when calculating the height of the object.]

Extension

Early in the school year, spend time creating "measuring tools," which can be used throughout the year. One handy tool could be made of heavy string onto which students have marked off meter lengths (30-50 meters should be long enough). The marks need to be done with dark, bold permanent markers. Then have them label every five meters, using masking tape and indelible markers. Wrap the strings around a box, or can, or large spool for storage.

Constructing Your Clinometer

1. You will need a copy of the clinometer pattern on card stock, a 20 cm piece of string, a paper clip, a ruler, and transparent tape to construct your clinometer.
2. Poke a hole at the spot labeled *Plumb Bob Anchor*. Thread the string through this hole so that about 2 cm hangs on the backside of the clinometer. Tape this short end securely to the back.
3. Tie a paper clip to the other end of the string and make sure that it swings freely from the vertex.
4. To provide a sighting plane, tape the broad upper band of the clinometer to your ruler.

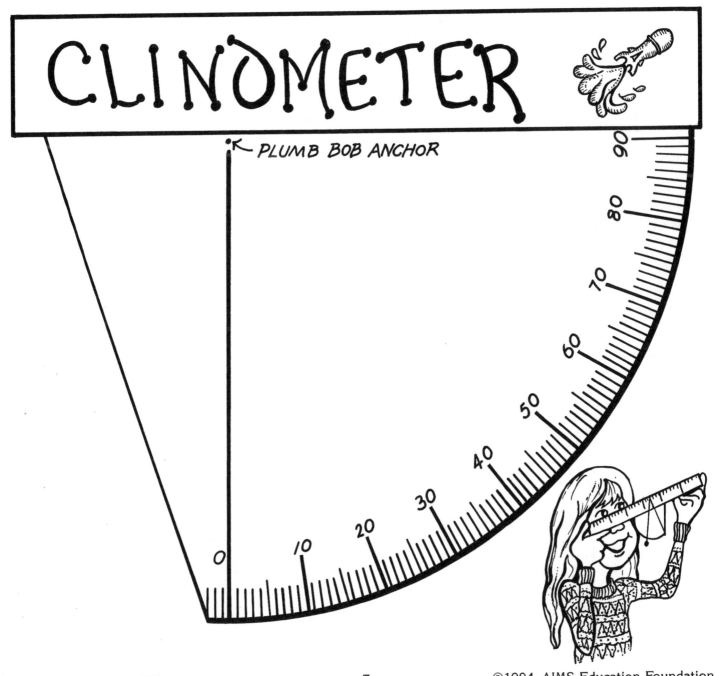

CLINOMETER

← PLUMB BOB ANCHOR

90 80 70 60 50 40 30 20 10 0

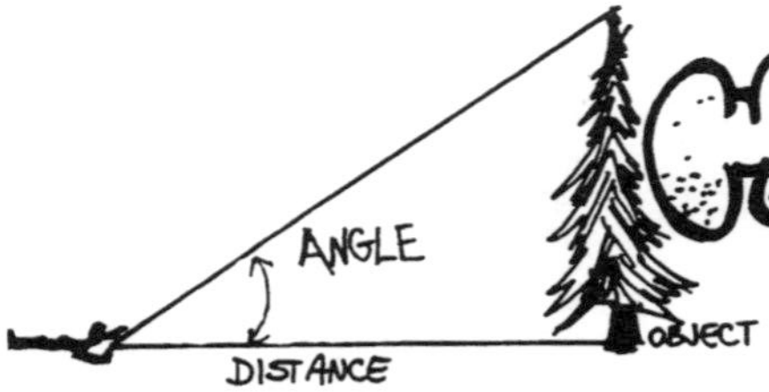

HOW HIGH IS IT?

NAME OF OBJECT	ESTIMATE OF OBJECT'S HEIGHT	ANGLE	TANGENT OF ANGLE	X DISTANCE FROM OBJECT	= HEIGHT OF OBJECT
1.				m	m
2.				m	m
3.				m	m
4.				m	m
5.				m	m
6.				m	m
7.				m	m
8.				m	m
9.				m	m
10.				m	m
11.				m	m
12.				m	m
13.				m	m
14.				m	m
15.				m	m
16.					m

ANGLE
DISTANCE
OBJECT

THE SKY'S THE LIMIT!　　　8

HOW HIGH CAN YOU THROW?

Topic
Determining heights of objects

Key Question
How high can you throw?

Focus
Students will determine the amount of time it takes a thrown ball to hit the ground.

Math
NCTM Standards
- *Generalize solutions and strategies to new problem situations*
- *Acquire confidence in using mathematics meaningfully*
- *Extend their understanding of measurement*
- *Estimate, make, and use measurements to describe and compare phenomena*

Measurement
  altitude
  time
Using formulas

Science
Project 2061 Benchmarks
- *Results of scientific investigations are seldom exactly the same, but if the differences are large, it is important to try to figure out why. One reason for following directions carefully and for keeping records of one's work is to provide information on which might have caused the differences.*
- *Measuring instruments can be used to gather accurate information for making scientific comparisons of objects and events and for designing and constructing things that will work properly.*
- *Measurements are always likely to give slightly different numbers, even if what is being measured stays the same.*
- *Tables and graphs can show how values of one quantity are related to values of another.*

Physical science
  gravity

Processes
Observing
Collecting and recording data
Interpreting data
Applying and generalizing

Materials
Stopwatch
Ball or other object to throw into the air

Background Information
In this investigation, students will determine the height of a thrown object using time and the laws of gravity as the tools for measurement. While an object is in the air, half the time it is rising and half the time it is falling back to earth. During the first half, the force of gravity is slowing the ball at the same rate that it will speed up the ball as it returns to earth.

Timing with a stopwatch is a useful mathematical, scientific, and real-life skill which many students have not had an opportunity to develop. Because stopwatches are used so often in these activities, try to obtain several for the students to use. Allow students time to practice, preferably near the beginning of the year; you might have them time a number of simple, everyday school tasks such as getting in line for lunch, taking out their books and finding the right page, and so forth. You might even end up with a classroom of students who are quick and efficient!

Management
1. This will take a minimum of two 45-60 minute periods.
2. The activity can be a large group investigation but works best in groups of three to six.
3. Warn students to never throw sharp or dangerous objects. Each person should pay strict attention to the flying object(s) to prevent possible injury.
4. You may want to assign specific jobs to each student in a group, i.e. Timer, Recorder, Thrower, Observer, etc.
5. **Caution:** Students working on this investigation should pick an open area, away from crowds.
6. For younger students, multiply by 5 meters instead of multiplying by 490 cm and dividing by 100.

Procedure

1. Select an area away from buildings and people.
2. The timer will make ready with the stopwatch. At the timer's command, the thrower will throw the object into the air.
3. The timer will time the throw from the moment the ball leaves the thrower's hand until it hits the ground.
4. The timer will announce the time of the throw, which is then recorded on the data sheet.
5. Repeat as necessary to complete data sheet.
6. Do calculations to determine distances.

Extensions

1. You may extend this investigation by changing the shape or size of the object.
2. A clinometer may be used to determine the altitude.

How High Can You Throw?

NAME _____

	UP + DOWN TIME	DOWN TIME (÷2)	DOWN x DOWN TIME x TIME	x 490cm	÷ 100 = DISTANCE
FIRST THROW				cm	m
SECOND THROW				cm	m
THIRD THROW				cm	m
TOTAL			TOTAL		m
÷ 3 = AVERAGE			÷ 3 = AVERAGE		m

AVERAGE TIME = _____ SECONDS

AVERAGE DISTANCE = _____ meters

BEST ☆ DISTANCE _____ meters

UNBELIEVABLE FLYING OBJECTS

Topic
Geometric constructions and flying shapes

Key Question
What effect does the shape, number of blades, size, or material of construction have on the flight of a spinning object?

Focus
Students will construct various symmetric shapes using rulers, compasses, and protractors and then explore their flight properties.

Math
NCTM Standards
- *Extend their understanding of the concepts of perimeter, area, volume, angle measure, capacity, and weight and mass*
- *Represent and solve problems using geometric models*

Measuring
    length
    angles
Geometry
Using tools
    ruler
    protractor
    compass

Science
Project 2061 Benchmarks
- *Doing science involves many different kinds of work and engages men and women of all ages and backgrounds.*
- *Many objects can be described in terms of simple plane figures and solids. Shapes can be compared in terms of concepts such as parallel and perpendicular, congruence and similarity, and symmetry. Symmetry can be found by reflection, turns, or slides.*

Processes
Observing
Predicting
Identifying and controlling variables
Generalizing

Materials
Rulers
Compasses
Protractors
Scissors
6" (15 cm) squares of tag board (enough for 3 per student)

Background Information
An object with point symmetry will be balanced around a central point. Air makes the spinning object tilt upward as it moves along. Then, it begins to slip on the air and, if it has point symmetry, come back quite close to the thrower like a boomerang.

Management
1. To do this activity, allow anywhere from 20 to 50 minutes, depending on the ability of students to use rulers, compasses, and protractors.
2. Proper use of rulers, compasses and protractors should be taught prior to this activity.
3. Have students study *Figures 1, 2,* and *3* on the activity page to get an idea of what their UFO's will look like.
4. These shapes can be constructed on different days, or all together on one day using tagboard, old file folders, etc.
5. The shapes can be constructed as a whole class activity with the teacher using an overhead projector. Step-by-step construction procedures can be demonstrated using a transparent ruler, protractor, and compass with an eraser on its point.
6. Younger students can be given the shapes which have been copied onto tagboard; then they simply cut them out and fly them.

Procedure
1. For step-by-step construction information, see the following page. Construct the first shape on a 6-inch (15 cm) square piece of tagboard.
2. Cut out the shape and place it on a book with one blade extending over the edge (see diagram on student page).
3. Tilt the book slightly upward and give the blade a sharp rap with a pencil or your finger to impart a spinning motion.
4. Observe what happens and predict how the second shape will fly.

5. Construct the second shape and launch it. Were your predictions correct?
6. Construct and fly the third shape.
7. Based on the experiences gained by building and flying the first three UFO's, some students may be ready to construct more shapes of different dimensions, shapes, number of blades or construction material. Allow students time to construct their own UFO's or manipulate the variables of the ones they have already constructed.
8. Have them predict and test the effects of these changes (variables)on the flight of their UFO's.

Discussion

1. Why do the UFO's fly? [The spinning blades create lift, just like the wing of a plane or a helicopter's rotor.]
2. Does the angle of launch affect the flight path of the UFO?
3. What is the flight path when the UFO is spinning counterclockwise?...clockwise? Why?
4. What effect does changing the number of blades, size, shape, etc., have on the UFO?

Extensions

1. Have students color the UFO blades and observe the color-mixing effect. (This is similar to the spinning color wheel.)
2. Try and find the shape that boomerangs best.
3. Have students identify blade angles as acute, right, or obtuse.
4. Have students find and mark lines and points of symmetry on their UFO's.
5. Have students write and/or illustrate the step-by-step construction procedure of either a given pattern or their own design.

Tilt book slightly then give UFO blade a sharp rap with your finger!

Construction of UFO shown in *Figure 1*:
A ruler construction

1. Find the center of each side of the 6-inch (15 cm) square piece of tagboard, and mark.
2. Draw two perpendicular lines connecting the marks on opposite sides.
3. Draw parallel lines on either side of the two original line, 1 cm away, forming blades 2 cm wide.

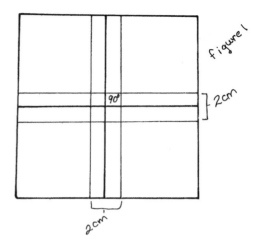

• •

Construction of UFO shown in *Figure 2*:
A compass-and-ruler construction

1. Spread the point and pencil end of a compass 7 cm apart using the ruler as a measure.
2. Find the center of a tagboard square and put the point of the compass there. Construct a circle with a radius of 7 cm.
3. Make a mark anywhere on the circle. Using the radius of the circle as a measure, put the point of the compass on the first mark and make another mark on the circle with the compass. Move the compass point to the new mark and repeat the process until you have six marks on the circle.
4. Draw a line from the center of the circle to one of the marks on the circle. Going clockwise from that line, skip one mark and draw a line from the center of the circle to the next mark. Continue clockwise from this new line skipping one mark and drawing your final line. You now have three lines about 120° apart.
5. Make three blades 2 cm wide by constructing parallel lines on both sides of the three original lines, 1 cm away.

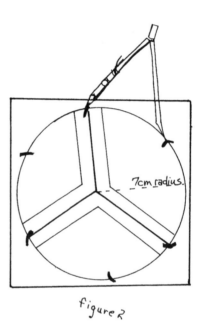

• •

Construction of UFO shown in *Figure 3*:
A protractor-and-ruler construction

1. Find the center of the top edge of the square and mark it.
2. Draw a line 7.5 cm long straight down from the mark, and make a dot at the end.
3. Place the protractor on this line so that its center point is on the dot and its left-side baseline is on top of the line you drew.
4. Find the 120° mark on the protractor and make a dot.
5. Connect this dot with a 7.5 cm line running to the center dot.
6. Rotate the protractor so that its center point is on the center dot and its right-side baseline is on the first line you drew.
7. Find the 120° mark on the protractor and make another dot.
8. Connect this dot with a 7.5 cm line running to the center dot.
9. Draw parallel lines on both sides of the three original lines, making blades 2 cm wide.

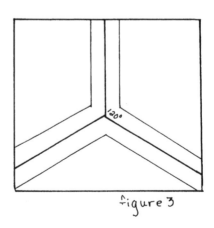

THE SKY'S THE LIMIT!   14   ©1994 AIMS Education Foundation

Unbelievable Flying Objects

Use a 7cm radius to draw your circle. Keep the compass at 7cm and mark off 6 equal arcs.

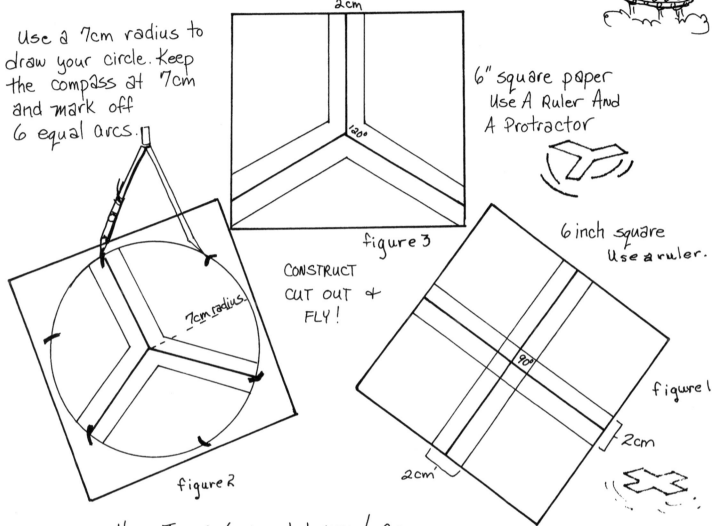

2cm

120°

figure 3

6" square paper
Use A Ruler And
A Protractor

6 inch square
Use a ruler.

CONSTRUCT
CUT OUT &
FLY!

7cm radius.

90°

figure 1

2cm

2cm

figure 2

Now Try a 6 pointed UFO! Can you make one with 5 sides! Construct your own UFO of a different size and shape.

Tilt Book Slightly

Give UFO Blade a sharp rap with your finger!

Sketch all the UFOs that will boomerang!

More Unbelievable Flying Objects

Topic
Line drawings and flying shapes

Key Question
What do you predict will be the flight characteristics of the shapes you construct?

Focus
The students will construct several line drawings and explore the flight characteristics of the resulting shapes.

Math
NCTM Standards
- *Extend their understanding of the concepts of perimeter, area, volume, angle measure, capacity, and weight and mass*
- *Represent and solve problems using geometric models*

Geometry
Symmetry

Science
Project 2061 Benchmarks
- *Learning means using what one already knows to make sense out of new experiences or information, not just storing the new information in one's head.*
- *Many objects can be described in terms of simple plane figures and solids. Shapes can be compared in terms of concepts such as parallel and perpendicular, congruence and similarity, and symmetry. Symmetry can be found by reflection, turns, or slides.*

Processes
Observing
Predicting
Generalizing

Materials
Tagboard or other similar material
Rulers
Scissors

Background Information
1. Line drawings use pairs of points connected by line segments to approximate a curve.
2. If the sets of dots converge in an angle, the drawing will by symmetrical around the bisector of that angle (*Figure 1*). Drawings can also have symmetry (*Figure 2*).

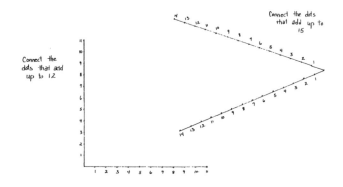

Figure 1

3. For further inspiration, please see the listings in the *Bibliography* at the end of this volume.

Management
1. This activity takes approximately 30-40 minutes.
2. The first student page should be done as a practice activity before working on the second.
3. Completed line drawings may be used as examples to help students visualize the process.
4. Young students may be started on line drawings by using a geoboard and rubber bands.

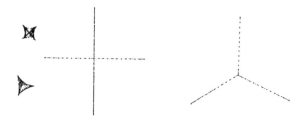

Figure 2

Procedure

1. Have students work with geoboards or do the first student page to become familiar with the line drawing process.
2. If possible, have the second student page run off on tagboard. If this is not possible, the drawing can be done on regular paper and the outline traced onto tag afterwards.
3. Using a completed line drawing as an example, have students do the line drawings.
4. Cut out the line drawing if it is on tag board; otherwise, trace its outline onto the tag and cut it out.
5. Have students predict the flight characteristics of the line drawings based on their experiences with flying shapes in the previous activity (*Unbelievable Flying Objects*).
6. Fly the line drawings using the same procedure as in *Unbelievable Flying Objects*.
7. Observe and see if the predictions were correct.

Discussion

1. Follow the same general questions as you covered in *Unbelievable Flying Objects*.
2. Where are the lines of symmetry in the two drawings? Where is the point of symmetry?
3. Is it possible to make an asymmetrical shape which will fly? Try it.
4. Is it possible to make a shape which will fly which just has line symmetry? Try this, too.

Extensions

1. Have students create their own line drawings.
2. Color in the line drawings and create new ones as an art activity.
3. Try to make much larger drawings and see whether they fly.
4. Use thread and needles on cardboard to make string art.
5. Try to fly some of the line drawing shapes that don't have symmetry.

Line Drawings

NAME

Connect the dots that add up to 12.

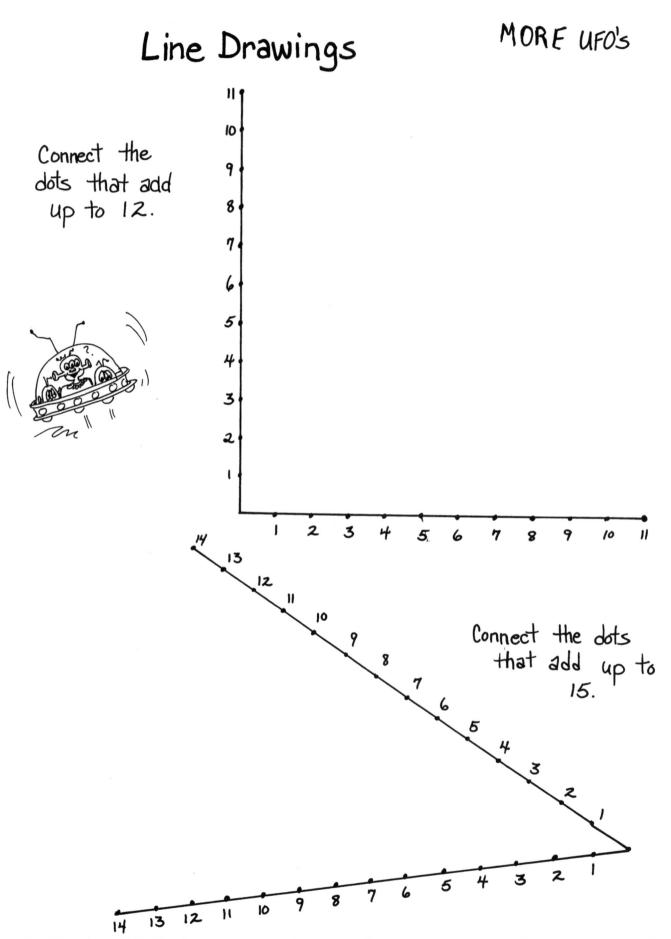

Connect the dots that add up to 15.

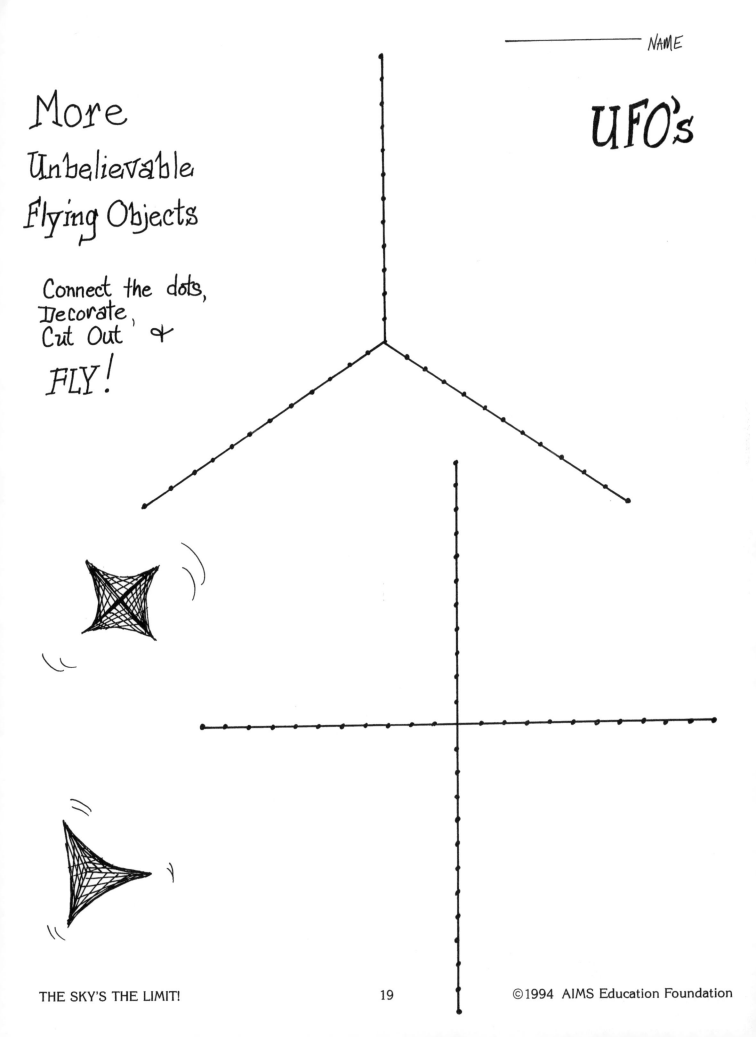

More
Unbelievable
Flying Objects

UFO's

Connect the dots,
Decorate,
Cut Out &
FLY!

.... But Will It FLY ?

Topic
Polygons

Key Question
Can you construct a polygon that will fly using only popsicle sticks?

Focus
Using popsicle sticks as a construction material, students will explore the characteristics of a variety of polygons.

Math
NCTM Standards
- *Identify, describe, compare, and classify geometric figures*
- *Represent and solve problems using geometric models*

Geometry
Measuring
  angles
Using formulas

Science
Project 2061 Benchmarks
- *Learning means using what one already knows to make sense out of new experiences or information, not just storing the new information in one's head.*
- *There is no perfect design. Designs that are best in one respect (safety or ease of use, for example) may be inferior in other ways (cost or appearance). Usually some features must be sacrificed to get others. How such trade-offs are received depends upon which features are emphasized and which are down-played.*

Processes
Observing
Collecting and recording data
Interpreting data
Generalizing
Communicating

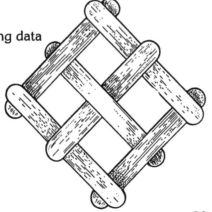

Materials
Per student:
  20-25 popsicle or craft sticks
  drawing paper or graph paper

Background Information
  This activity can involve very little, or a great deal of technical information, depending on the grade level of your students, the time available, and your expectations. (Students can handle a great deal of apparently complex material if they discover it themselves.) The basic goal of the investigation is to acquaint students with the names and basic features of "standard" polygons. Some terms which may be necessary or helpful in your discussions are listed below. More detailed, advanced information can be found in the *Supplementary Information* which follows this section.

1. polygon – a many-sided, geometric feature

2. closed fixture – all sides meet to enclose a finite space

3. regular polygon – all sides and angles are congruent, or equal in measure

4. vertex – point at which two sides of a polygon, rays of an angle meet

5. acute angle – measures less than 90°, but more than 0°

6. obtuse angle – measures more than 90°, but less than 180°

7. right angle – measures 90° (made by perpendicular lines)

8. triangle-3 sides, 3 angles

9. quadrilateral -4 sides, angles

10. pentagon – 5 sides, angles

11. hexagon – 6 sides, angles

12. octagon – 8 sides, angles (a stop sign)

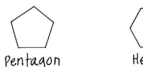 

Pentagon   Hexagon   Octagon

Management

1. Allow 30 minutes for initial exploration. Then allow 40-60 minutes for the student activity on *Sheet 1*. Other student activities, *Sheets 2* and *3*, will also take 40-60 minutes, if you decide to do them.
2. Divide the class into pairs from the beginning. Each student will construct his/her own polygons, but testing will be easier in pairs.
3. To make construction of polygons easier, soak the popsicle sticks in water for about 15 minutes.
4. Use a large, open area for constructing polygons – the sticks will tend to pop out at first. Warn students to be careful.
5. Allow at least one day, or period, for free exploration and experimentation. There is skill involved in making these, and practice will help.
6. Emphasize that students cannot use tape or glue, but they must be able to pick up the figures they make.
7. When you are ready to fly the polygons, use a large area as they will probably fall apart, and you don't want students to be hit with flying sticks.
8. Lay out a line from which to launch polygons, so it will be easy to measure and compare the distance each travels.
9. As student begin experimenting, answer questions that come up, but do not feel that you have to offer elaborate explanations or detailed information. Students do not have to know everything – names, e.g., – before they begin.
10. Have students work through *Sheet 1*, drawing their polygons. Then, worry about whether the students can name the figure they have made.
11. *Sheets 2* and *3* can be used if you feel your class can handle them, or a few, interested students could work on them independently.

Procedure

1. On the first day, tell students that a polygon is a "figure with many sides, " and that they are going to build polygons using popsicle sticks. Have students volunteer to draw examples on the board. Discuss what makes a figure a polygon – a dog isn't; a triangle is – and eliminate those which are not closed, geometric figures.
2. Have students construct polygons with popsicle sticks and sketch them on *Sheet 1*.
3. Help them to learn to identify the figures by their geometrical names and how many sides and angles each figure has.
4. Investigate the angles of polygons – size, type (*Sheet 2*).
5. Fly the polygons and record data including the distance of the flight (*Sheet 3*).

Discussion

1. For the most part, discussion should be guided by student questions and comments.
2. Most areas of discussion are documented on student pages.

Extensions

1. Construct a quadrilateral as before. How many other figures can you name be adding our shapes to it? Sketch them. Count and record the number of sides and braces needed. Test them in flight. What effect do the additions have?
2. Determine the areas of your polygons by sketching them on graph paper and by using available formulas.
3. Do convex or concave polygons fly better? Test your hypothesis.
4. Do symmetrical or asymmetrical polygons fly better? Test.
5. Is there a way, still without glue, of increasing the stability by doubling the layers of sticks?
6. What happens if you do glue or lash the sticks? Which one flies better?
7. Will covering the frame of the polygon make it fly better?
8. Try the same activity using tongue depressors. Does it work?

● ●

Supplementary Background Information
...But Will It Fly?

Identification
1. **Polygons** are many-sides figures. Regular polygons have all sides and angles equal ($\cong$ congruent).
2. **Triangle:** 3 sides; sum of interior angles = 180°; no diagonals. **Equilateral Triangle:** equal sides and equal angles; all angels = 60°.
3. **Quadrilateral:** 4 sides; sum interior angles equals 360°; 2 diagonals.
   a. **Parallelogram:** opposite angles equal; opposite sides parallel (| |) and equal.
   b. **Rectangle:** 4 right angles; opposite sides congruent and parallel.
   c. **Square:** 4 90° angles; all sides equal; opposite sides parallel.
   d. **Rhombus:** 4 equal sides; opposite angles equal.
4. **Pentagon:** 5 sides; sum of exterior angles = 360°; 3 diagonals. In regular pentagon, interior angles are 108°; exterior angles measure 72°.
5. **Hexagon:** 6 sides; sum of exterior angles equals 360°; 4 diagonals. In regular hexagon, interior angles each measure 120°, exterior angles measure 60°.
6. **Octagon:** 8 sides; sum of exterior angles = 360°; 6 diagonals. In regular octagon. interior angles each measure 135°; exterior angles measure 45°.

Formulas
Angles and other information of a polygon of n sides,
   a. n-2 = number of diagonals
   b. $\frac{360}{n}$ = measurement of an exterior angle
   c. $\frac{180(n-2)}{n}$ = measurement of an interior angle in a regular (or 180 - measurement of exterior angle)
   d. $\frac{360}{n}$ = measurement of central angle

Areas: B = base, h = height or altitude, s = side
   a. Triangle: A = 1/2 Bh
   b. Rectangle: A = Bh (1w)
   c. Parallelogram: A = Bh (However, stress the fact that the altitude is not congruent to the length of the side.)
   d. Square: A = s²

Miscellaneous
1. congruent $\cong$; parallel | |; perpendicular $\perp$; right angle ∟ ; angle $\angle$; m$\angle$a = measurement of angle a.

2.

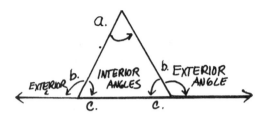

3. A polygon is named by its vertices. Thus, this figure is ABCDEF.

4. This is a concave polygon: This is a convex polygon:

.....But Will It FLY?

FLYING POLYGONS - SHEET #1

1. WHAT IS THE SMALLEST NUMBER OF STICKS YOU CAN USE TO BUILD A POLYGON WHICH YOU CAN PICK UP? _____

SKETCH YOUR POLYGON —

   A. HOW MANY SIDES DOES IT HAVE? _____ HOW MANY ANGLES? _____

   B. WHAT IS IT CALLED? _____

2. WHAT IS THE GREATEST NUMBER OF STICKS YOU CAN USE TO CONSTRUCT A POLYGON YOU CAN PICK UP? _____

SKETCH YOUR POLYGON —

   A. HOW MANY SIDES DOES IT HAVE? _____ HOW MANY ANGLES? _____

   B. DOES IT HAVE A NAME? _____

3. SKETCH ANY OF THESE POLYGONS WHICH YOU CAN CONSTRUCT.
   A. 3 SIDES   B. 4 SIDES   C. 5 SIDES   D. 6 SIDES

A. POLYGON PARTICULARS

HOW MANY BRACES DID YOU NEED?

WHAT TYPE OF ANGLES ARE IN YOUR POLYGONS?

ACUTE <90°   90° RIGHT (ON!)   OBTUSE >90°

# OF SIDES	# OF BRACES	# OF ANGLES	TYPE OF ANGLES	NAME OF POLYGON
3				
4				
5				
6				
7				
8				

B. ULTIMATE ANGLES – FOR THE BRAVE AND CLEVER!

DRAW, OR TRACE, AROUND THE OUTSIDES OF THE POLYGONS YOU HAVE MADE.

1) MEASURE THE INTERIOR ANGLES. FILL IN THE TABLE BELOW.

2) MEASURE THE EXTERIOR ANGLES. ADD THE INFORMATION TO YOUR TABLE.

EXTERIOR   INTERIOR ANGLES   EXTERIOR ANGLE

# OF SIDES	# OF ANGLES	MEASUREMENT – INTERIOR ∠'s	MEASUREMENT – EXTERIOR ∠'s
3			
4			
5			
6			
7			
8			

DO YOU THINK THERE IS A PATTERN?

CAN YOU WRITE A FORMULA FOR FINDING THE MEASUREMENT OF AN EXTERIOR ANGLE?

CAN YOU WRITE A FORMULA FOR FINDING THE MEASUREMENT OF AN INTERIOR ANGLE?

....But Will It "FLY"?

FLYING POLYGONS - STUDENT SHEET 3

NOW THAT YOU HAVE CONSTRUCTED THE POLYGONS,
   SEE IF THEY WILL FLY!
WORK WITH YOUR PARTNER SO THAT YOU HAVE 2
   OF EACH TYPE TO TEST.

A. HOW FAR CAN YOU FLY EACH OF YOUR POLYGONS?
   1. MEASURE THE DISTANCE FROM THE LAUNCH LINE TO THE POINT AT WHICH EACH HIT
      THE GROUND.
   2. RECORD THE RESULTS HERE. HAVE A CLASS CHART.

TYPE OF POLYGON	DISTANCE	TYPE OF POLYGON	DISTANCE
TRIANGLE A		HEXAGON A	
TRIANGLE B		HEXAGON B	
SQUARE A			
SQUARE B			
PENTAGON A			
PENTAGON B			

   WERE THE RESULTS CONSISTENT AMONG GROUPS?
   DO CERTAIN POLYGONS SEEM MORE STABLE?

B. AND WHEN THE WHOLE SYSTEM FALLS APART . . . . .
       (WHICH IT PROBABLY WILL)
   HOW FAR APART DID THE STICKS LAND WHEN THEY HIT THE GROUND?
   1. WHEN EACH POLYGON HITS THE GROUND AND FALLS APART, <u>CIRCLE</u> THE AREA WITH A
      LENGTH OF STRING WHICH ENCLOSES <u>ALL</u> THE STICKS.
   2. MEASURE THE LENGTH OF THE STRING NEEDED (CIRCUMFERENCE OF THE CIRCLE,
      $C = 2\pi r$) FIND THE AREA OF THE CIRCLE ($A = \pi r^2$) $\pi = 3.14$ or $\frac{22}{7}$
   3. MAKE A CHART FINDING YOUR DATA. DOES THE AREA OF THE CIRCLE
      COVERED INCREASE AS THE NUMBER OF STICKS INCREASE?
   4. DOES THE AREA OF THE CIRCLE INCREASE CONSISTENTLY AS THE
      DISTANCE FLOWN INCREASES?

                                    OR, PLAY "PICK UP STICKS"

THE SKY'S THE LIMIT!   25   ©1994 AIMS Education Foundation

Topic
Paper rotors, twisters

Key Questions
1. How does a paper rotor or twister behave?
2. How can the behavior be altered?

Focus
Students will investigate the behavior of a paper rotor, testing for both the rate and accuracy of the fall, and will examine ways in which the behavior can be altered.

Math
NCTM Standards
- *Systematically collect, organize, and describe data*
- *Analyze tables and graphs to identify properties and relationships*
- *Develop the concepts of rates and other derived and indirect measurements*

Measurement
   length
   time
Using formulas

Science
Project 2061 Benchmarks
- *Offer reasons for their findings and consider reasons suggested by others.*
- *The earth's gravity pulls any object toward it without touching it.*
- *Measuring instruments can be used to gather accurate information for making scientific comparisons of objects and events and for designing and constructing things that will work properly.*

Physical science
   forces affecting flight

Processes
Predicting
Observing
Manipulating variables
Collecting and recording data
Generalizing

Materials
Rotor patterns copied on paper
Other types of paper-file cards, construction paper, etc.
Scissors
Metric tapes
Stopwatches – one for each group of students, if possible
Targets

Background Information
Helicopters are really airplanes with moving wings in which the usual airplane wing and propeller are replaced by a large horizontal rotor. The helicopter rises because the rotor blades push down against the air.

As these paper models fall they will spin, imitating the rotation of the rotor blades of a real helicopter. They will not rise since there is no input of power, but the spin will reduce the rate of fall because it increases the lift.

The behavior of the twisters can be altered by changing: (1) the type of paper used; (2) the size of the twister made using the same proportions; (3) the position of the rotor blades (the pitch, or the angle in the relation to the shaft); (4) the design of the twister; and (5) the height from which the twisters are dropped.

In testing their twisters, students will be able to compute the rate of fall in centimeters per second using the formula $S = r \cdot t$, or $r = d/t$, where S and d stand for the distance of the fall, t is the time in seconds, and r is the rate of fall.

Management
1. This investigation breaks down into three logical sections: (a) construction of the two basic twister designs, for which patterns are given; (b) testing of the two models for the rate of fall and the accuracy of the fall; and (c) construction and testing of individual designs in which students change the independent variables.
2. The activity can be done in 2 or 3 class periods of about 40 minutes each.
3. For all tests, the height from which the twisters are dropped should remain constant. 150 cm is a convenient height. However, you may find it possible and desirable to use a greater height, distance of fall.
4. If students work in pairs, one can make each of the versions for which patterns are given, and they can compare the results of their tests using the activity sheet.

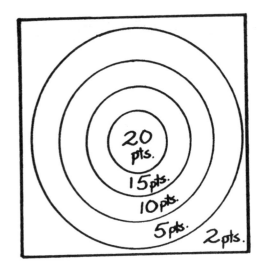

5. If you wish, two, three, or four pairs of students can be grouped together to use one stopwatch and one target.
6. You may want to have a supply of paper clips available in case the twisters seem to need added weight at the bottom of the shaft.
7. Targets can be easily made by a few students. Use a piece of poster board (22 x 28"); draw four concentric circles (r = 2", 4", 7", 10"); add point values. The targets can be painted and laminated for future use.

Procedure
1. Construct the targets.
2. Using the patterns provided, construct the two versions of the twister. Have students identify which pattern each will use by naming the "helicopters."
3. Determine heights from which copters are to be dropped. Find a way to insure that this is consistent, and not influenced by the height of the student.
4. Time the fall of the two models using a stopwatch. Record the data. Compute the rate of fall in centimeters per second using the formula $S = r \cdot t$ or $r = d/t$.
5. Test the models for accuracy using a target placed on the floor and tallying points.
6. After discussing the behavior of the twisters, design and construct other twisters using the materials, size and proportions, rotor position, and design of their choice.
7. Test the newly constructed models against the original models and record data.

Discussion
1. How does the twister behave? (You might drop a model in which the rotor blades have not been bent back to stimulate discussion.)
2. Which version is best? What constitutes "best?"
3. Do the twisters always rotate in the same direction? (Try bending the rotors and flaps in different directions.)
4. What happens if you change the angle at which the blades are bent?
5. What happens if you shorten the blades?
6. What will happen if you change:
    a. the size;
    b. the proportions;
    c. the material used?

Extensions
1. Test identical models dropped from different heights. Graph the results.
2. Test identical models made of different materials. Graph the results.
3. Construct a class graph for each of the given versions, which shows the rates at which they fall.
4. Have a class contest to see who can make the most accurate helicopter.
5. Research the history of helicopters. Find out about the various uses in today's world.

BE A ROTOR PROMOTER

Adopt A Copter Today!

HERE ARE TWO PATTERNS FOR A PAPER ROTOR OR TWISTER - A HELICOPTER. CHOOSE A PARTNER: IF ONE OF YOU CONSTRUCTS THE ALPHA AND ONE THE BETA, YOU CAN COMPARE THE PERFORMANCES.

TRY YOUR MODELS MADE OUT OF FILING CARDS (5x8) INSTEAD OF PAPER. TRY OTHER MATERIALS.

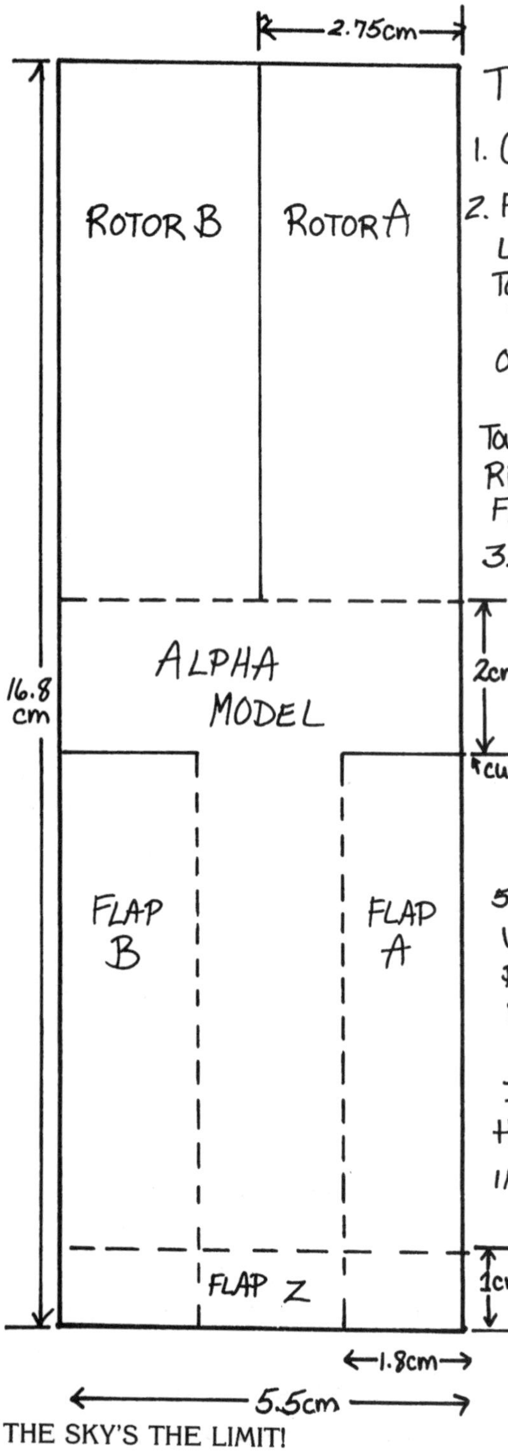

ALPHA MODEL

|← 2.75cm →|

ROTOR B | ROTOR A

ALPHA MODEL

FLAP B | FLAP A

FLAP Z

16.8 cm

2cm

CUT

1cm

←1.8cm→

← 5.5cm →

To BUILD:

1. CUT ALONG THE SOLID LINES.

2. FOLD ALONG THE DASHED LINES. FIRST, FOLD FLAP A IN TOWARD THE CENTER.
   THEN, FOLD FLAP B IN OVER FLAP A.
   NEXT, FOLD FLAP Z UP TOWARD YOU UNTIL IT MAKES A RIGHT ANGLE WITH THE OTHER FLAPS.

3. NOW, FOLD ROTOR A BACK AND ROTOR B FORWARD

4. DROP YOUR HELICOPTER IN A TEST FLIGHT TO DETERMINE WHETHER IT ROTATES.
   EXPERIMENT WITH A VARIETY OF ADJUSTMENTS.

5. WHEN YOU ARE SATISFIED WITH YOUR TEST FLIGHT, PROCEED TO THE RECORD/COMPETITION SHEET.

INCIDENTALLY, DOES YOUR HELICOPTER ALWAYS SPIN IN THE SAME DIRECTION?

CLOCKWISE OR COUNTERCLOCKWISE.

CAN YOU MAKE IT REVERSE???

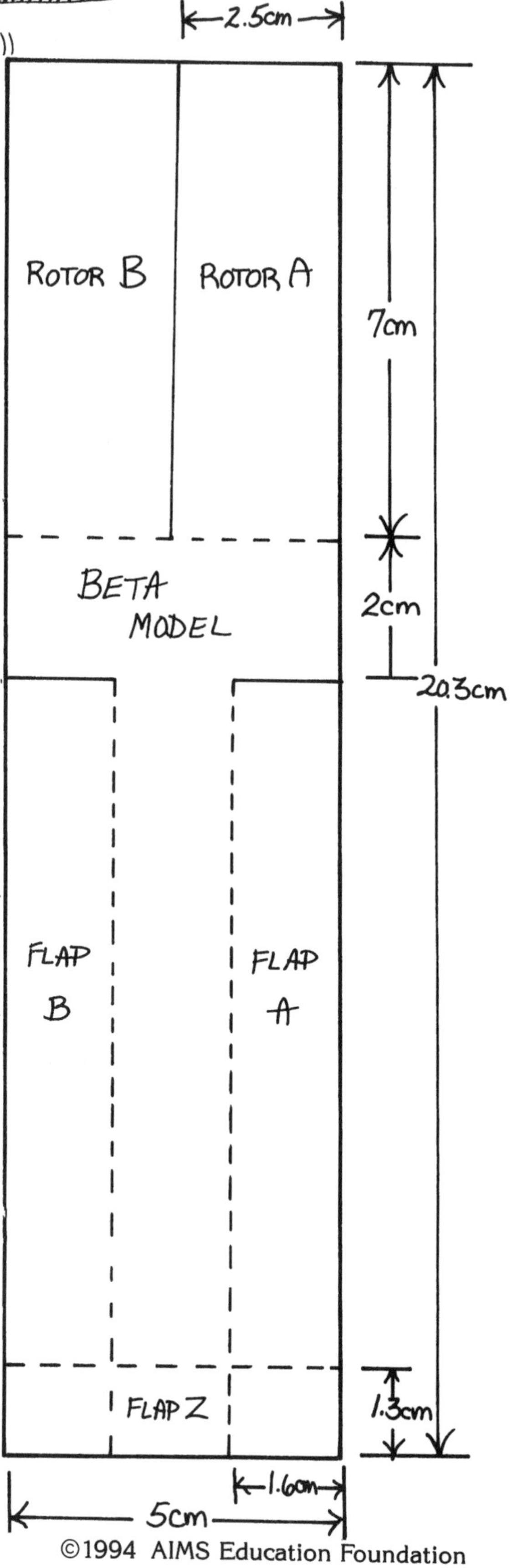

BETA MODEL

|← 2.5cm →|

ROTOR B | ROTOR A

BETA MODEL

FLAP B | FLAP A

FLAP Z

7cm

2cm

20.3cm

1.3cm

←1.6cm→

← 5cm →

BE A ROTOR PROMOTER

ADOPT A
COPTER TODAY!

RECORD/COMPETITION SHEET

MY HELICOPTER IS _____ MODEL

DISTANCE OF DROP _____ cm

DESCENT TIME

NAME _____ TIME

NAME _____ TIME

	TIME		TIME
TRIAL # 1		TRIAL #1	
TRIAL # 2		TRIAL # 2	
TRIAL # 3		TRIAL # 3	
TRIAL # 4		TRIAL # 4	
TRIAL # 5		TRIAL # 5	
TRIAL # 6		TRIAL # 6	
TOTAL		TOTAL	
AVERAGE		AVERAGE	

WHEN YOU HAVE FOUND THE TIME IN SECONDS, YOU CAN DETERMINE THE RATE AT WHICH YOUR HELICOPTER FALLS BY USING A FORMULA:

$$S = r \cdot t$$

S = DISTANCE (HEIGHT FROM WHICH YOU DROPPED YOUR TWISTER)

t = TIME IT TOOK TO HIT GROUND

YOU ARE TRYING TO FIND R.

WHAT CAN YOU DO IF YOU KNOW S AND t?

TRIAL # 1   r =

TRIAL # 2   r =

TRIAL # 3   r =

TRIAL # 4   r =

TRIAL # 5   r =

TRIAL # 6   r =

MAKE A DOUBLE BAR GRAPH COMPARING THE RATES OF YOUR COPTER WITH YOUR PARTNERS.

THE SKY'S THE LIMIT!

29

©1994 AIMS Education Foundation

BE A ROTOR PROMOTER

ADOPT A COPTER TODAY!

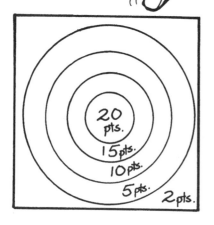

20 pts.
15 pts.
10 pts.
5 pts.   2 pts.

NOW THAT YOU HAVE FIGURED OUT THE RATE AT WHICH YOUR COPTER CAN FALL, SEE HOW ACCURATELY IT WILL LAND!

MAKE A POSTER BOARD TARGET AND PUT IT ON THE FLOOR. IN TURN, YOU AND YOUR PARTNER WILL DROP YOUR COPTERS FROM A GIVEN HEIGHT.

KEEP SCORE!

WHICH COPTER WAS MOST ACCURATE?

NAME	
TURN	SCORE
1	
2	
3	
4	
5	
6	
7	
8	
9	
10	
TOTAL	

NAME	
TURN	SCORE
1	
2	
3	
4	
5	
6	
7	
8	
9	
10	
TOTAL	

IT'S A REAL CORKER!

Topic
Air resistance

Key Question
In making a cork-and-feather helicopter, how will changing the number of rotor blades and the pitch of those blades affect the rate of fall?

Focus
Through experimentation, students will discover how changing the number of rotor blades and the angle at which they are placed affects the rate of fall of a helicopter device.

Math
NCTM Standards
  • *Make inferences and convincing arguments that are based on data analysis*
  • *Acquire confidence in using mathematics meaningfully*
Measuring
  angles
Geometry
Using a formula
Averaging

Science
Project 2061 Benchmarks
  • *If more than one variable changes at the same time in an experiment, the outcome of the experiment may not be clearly attributable to any one of the variables. It may not always be possible to prevent outside variables from influencing the outcome of an investigation (or even to identify all of the variables), but collaboration among investigators can often lead to research designs that are able to deal with such situations.*
  • *Measuring instruments can be used to gather accurate information for making scientific comparisons of objects and events and for designing and constructing things that will work properly.*

Processes
Observing
Collecting and organizing data
Interpreting data
Identifying and controlling variables
Generalizing

Materials
Corks:
   2-2 1/2" corks should be available from hobby shops, hardware stores, or scientific equipment supply houses. Wine bottle corks work well. (Make certain corks are not too dense.) Styrofoam spools from thread also work. Other alternatives: Styrofoam balls, eggs, or pieces.
Feathers:
   Have students collect feathers from the ground or purchase them from a hobby shop. Don't get them too big or too fluffy. Pieces from file cards and Styrofoam foam meat trays or fast-food containers can be pinned on in place of feathers.
Protractors
Stopwatches
Toothpicks

Background Information
   The principles behind the "corker-copter" are the same as those behind the paper helicopters. However, students will be manipulating two variables – the number of rotor blades used and the angle, or pitch, at which they are placed. If the corks (or whatever is used) are of the same size and weight, and if the measurements of the angles are made as accurately as possible, the results of the experiments in design will be more "honest" and more obvious.

Management
1. This activity will take two class periods of 40-60 minutes each. It would probably be best to allow some free investigation time before doing the formal activity, as students will love the fact that these creations spin around. They will want to know why the copters spin, make them spin faster, see them turn over when they are dropped bottom side up, and all sorts of other things before settling down to making them go slower.
2. This activity is divided into two parts: *Part A* has students construct copters with 2, 3, 4, or 5 rotor blades. They test the various designs, record data, average the times of the fall, and determine the rate

THE SKY'S THE LIMIT!

31

©1994 AIMS Education Foundation

of fall for each version. In *Part B,* students construct copters using the number of blades selected from *Part A;* then they adjust the angle or pitch of the blades. The designs are tested, data is recorded, the rate of fall is determined as above.

3. Students should construct their first copter, spend time observing and experimenting with it , and do the first activity *(A)* on one day, and the second activity *(B)* on the second day.
4. Working in groups of four makes both the construction and testing more efficient.
5. Have each member of the group make a different version of the copter. Students will mark their copters with their names, distinctive designs, or names of the "aircraft."
6. Emphasize the fact that the object of the activity is to maximize the length of time it takes for the copter to reach the ground. For once, the slowest person wins!
7. Be sure to drop the copters from a definite height such as the school stage, down a stairwell, or off the playground equipment if the air is calm. The copters behave somewhat erratically at first and need the distance to stabilize.
8. Have a model for the students to look at before they begin construction. Do not bring in an optimum design, however, or they won't make an effort.
9. To help in spacing the blades, evenly divide and mark off a short piece of string according to the desired number of blades. Wrap around the cork at desired distance from the top and mark appropriate positions. Use the toothpicks to poke the holes for the feather placement.
10. To measure the angles for *Part B,* it may be easier to use pre-measured and pre-cut pieces of tag board than to use a protractor directly.

Procedure

Part A

1. Have students predict whether 2, 3, 4 or 5 blades will create the best helicopter.
2. Once students have constructed their models (one of each type for each group), they will test their predictions by dropping each version from the same height. They will time the drop to the nearest tenth of a second and determine the rate of fall using an average of 3 trials.

Part B

1. Have students predict what angle the blades should be set at to produce the best fall.

Note: These are supplementary angles; they can be measured whichever way is easiest.

2. As before, students will produce corker-copters with blades set at different angles. Make sure that all students use the same number of blades and that each group has selected a variety of angles.

3. Test as before to determine the optimum pitch of the blades.
4. If time permits, have students combine the results of both experiments and make a final, definitive version.

Discussion

1 What is the best design for a corker-copter? Why?
2. What happens with other versions?
3. What are some other ways these could be changed?
4. What did you learn from this activity?

Extensions

1. If you wish, run a simple class competition at the end with all students dropping their favorite copter simultaneously from the same height. Or capitalize on the interest in the rotation and have a spin-off contest in which students count the number of rotations in a given distance.
2. Use different sizes and weights of corks. Use different rotor materials.
3. Make a class record and graph of test results.
4. Compete against a commercial shuttlecock!

ANGLE OR PITCH OF ROTOR

IT'S A REAL CORKER!

IN THIS INVESTIGATION, YOU ARE GOING TO MAKE ANOTHER TYPE OF HELICOPTER, USING A CORK AND FEATHERS (OR SOME MODERN EQUIVALENT). YOU MAY MAKE AS MANY VERSIONS AS YOU WISH, BUT CHANGE ONLY ONE VARIABLE AT A TIME: EITHER THE NUMBER OF FEATHERS YOU USE OR THE ANGLE AT WHICH THEY ARE SET IN THE CORK.

SUGGESTION: USE A *TOOTHPICK* TO POKE THE HOLE WHERE YOU WANT THE FEATHERS BEFORE YOU TRY TO INSERT THEM.

☆ WHAT IS THE BEST DESIGN OF A "CORKER-COPTER"?

TO FIND OUT, TEST THE RATE OF FALL OF THE VARIOUS MODELS YOU DESIGN.

A. ☆ HOW MANY ROTOR BLADES SHOULD YOUR "CORKER-COPTER" HAVE?

SET THE BLADES (FEATHERS) AT EVEN INTERVALS AROUND THE CORK AND AT THE SAME ANGLE EACH TIME. DROP YOUR COPTER FROM THE SAME HEIGHT FOR EACH TEST. TIME ITS FALL AND FIND THE RATE. REMEMBER- $S = r \cdot t$

HEIGHT OF DROP = _____ cm   ANGLE OF BLADES = _____

NUMBER OF BLADES	2	3	4	5
TRIAL # 1				
TRIAL # 2				
TRIAL # 3				
TOTAL TIME				
AVERAGE TIME				
RATE OF FALL				

WHICH OF YOUR MODELS SEEMED TO WORK BEST?

COMPARE YOUR RESULTS WITH YOUR CLASSMATES! WERE THEIR RESULTS THE SAME AS YOURS?

NOW THAT YOU HAVE DECIDED HOW MANY BLADES TO USE, YOU WILL NEED TO DECIDE AT WHAT ANGLE THEY SHOULD BE PLACED. CARRY ON!

IT'S A REAL CORKER?

B. AT WHAT ANGLE, OR PITCH, SHOULD YOU PLACE THE ROTORS (BLADES) OF YOUR CORKER COPTER?

BE AS ACCURATE AS YOU CAN WHEN MEASURING THE ANGLES AT WHICH THE FEATHERS ARE SET, THE "PITCH OF THE ROTORS". DROP YOUR CORKER-COPTER FROM THE SAME HEIGHT YOU USED IN PART A. TIME ITS FALL.... EACH ADJUSTMENT AND FIND THE RATE.

ANGLE OR PITCH OF ROTOR

HEIGHT OF DROP _____ cm   NUMBER OF BLADES _____

ANGLE OF BLADES				
TRIAL # 1				
TRIAL # 2				
TRIAL # 3				
TOTAL TIME				
AVERAGE TIME				
RATE OF FALL				

WHAT WAS THE OPTIMUM ANGLE FOR PLACING THE BLADES OF YOUR COPTER?

COMPARE YOUR CONCLUSION WITH YOUR CLASSMATES! WERE THEIR RESULTS THE SAME AS YOURS?

WHAT OTHER FACTORS MIGHT INFLUENCE THE RATE OF FALL OF YOUR CORKER-COPTER?

YOU MAY WANT TO TEST YOUR TEST COPTER DESIGN AGAINST THOSE OF YOUR CLASSMATES. OR YOU MAY WANT TO TRY MAKING A CLASS GRAPH SHOWING HOW MANY STUDENTS SELECTED WHICH NUMBER AND PITCH OF BLADES.

Topic
Parachutes

Key Question
What is the rate of descent of your parachute?

Focus
Students will construct various parachutes and calculate their rates of descent.

Math
NCTM Standards
- *Make inferences and convincing arguments that are based on data analysis*
- *Compute with whole numbers, fractions, decimals, integers, and rational numbers*
- *Select and use an appropriate method for computing from among mental arithmetic, paper-and-pencil, calculator, and computer methods*

Measuring
   time
   distance
Calculating rate of descent
Averaging

Science
Project 2061 Benchmarks
- *Measuring instruments can be used to gather accurate information for making scientific comparisons of objects and events and for designing and constructing things that will work properly.*
- *Use, interpret, and compare numbers in several equivalent forms such as integers, fractions, decimals, and percents.*
- *Learning means using what one already knows to make sense out of new experiences or information, not just storing the new information in one's head.*

Processes
Observing
Collecting and recording data
Interpreting data
Generalizing

Materials
Cloth or plastic (see *Management*)
String or thread
Transparent tape
Various weights (see *Management*)
Stopwatches
Meter tapes

Background Information
   A parachute is a simple canopy which traps and holds air, and gradually lets it escape to bring a payload (an object which is tied to its shroud lines) safely to earth.
   Since students will be calculating the rate of descent for the parachutes they construct, remember that $d = rt$. The students will be dividing the height from which the parachute is dropped by the time (to the nearest tenth of a second) it took to touch ground, to obtain a rate in meters per second.

Management
1. The construction times will vary widely with the age and ability of the students. The testing activity and follow-up will take 45-60 minutes. Total time: about 75-90 minutes.
2. Several days before you want to do this activity, encourage students to bring various types of materials to make a parachute, such as rags, plastic bags, paper, old handkerchiefs, etc. Also, encourage students to bring kite string or thread and nuts, bolts, washers, or fishing weights to use as payloads.
3. When enough materials have been collected for the activity, have students construct parachutes individually or in a small group. If pressed for time, materials and/or space, have students construct a parachute at home and bring it to school.
4. For younger students who have trouble tying knots, use plastic for the canopy and tape the shroud lines onto it.
5. Give students time to test their parachutes before any formal activities are done with them.
6. Find an area where students can drop their parachutes: a stairwell, off the school stage, from the monkey bars or slide, etc.

7. The formal testing would probably work better if it were done in small groups so that no one has to wait very long. Students will drop the chute, time the drop and record the information.

8. Calculators can be used to compute rate of descent since students will be dividing by a decimal fraction (1.8 seconds for instance).

Procedure
1. Make parachutes.
2. Test parachutes informally - (recess).
3. Choose a suitable drop site where three drop heights can be marked. Mark the heights 1, 2, and 3.
4. Alternate the jobs of dropping, timing, and recording.
5. Drop each chute twice from each height and record results. For a class chart, each student can choose the best (the slowest) of the two drops at each point and record it on a class chart. Remember: "He who lands last laughs longest."
6. Repeat process with other groups until all parachutes have been dropped. Complete the class chart by filling in the rates.

Discussion
1. What things affect the rate of descent? [material, size, payload, height of drop]
2. Did the rate of your chute change from one height to another? Why? [acceleration due to gravity]
3. Select the five slowest rates of descent and the five fastest from the class chart. Have those students display and describe the parachutes. Were there similarities? What can you conclude?
4. How would you modify your parachute to improve its performance.

Extension
Have a contest a week later for the slowest descent.

Ah CHUTE !

USE ANY MATERIAL YOU WISH TO MAKE A PARACHUTE. DROP IT FROM 3 DIFFERENT HEIGHTS. TIME ITS DESCENT AND COMPUTE YOUR PARACHUTE'S RATE OF DESCENT.

TIE STRING TO 4 CORNERS OF A SQUARE PIECE OF MATERIAL

AIR RESISTANCE SLOWS ITS FALL!!

HEIGHT ÷ TIME = RATE OF FALL

HEIGHT	TRIAL#	HEIGHT	TIME	RATE
#1	1	m		
	2	m		
#2	1	m		
	2	m		
#3	1	m		
	2	m		
		TOTAL		
		(÷6) AVERAGE		

THE PLEASURES OF PARACHUTING ARE PLENTIFUL!

TIE THE 4 STRINGS TO A SMALL NUT AND DROP!

RATE OF DESCENT = $\dfrac{\text{HEIGHT OF DROP IN METERS}}{\text{TIME OF DROP IN SECONDS}}$

$$r = \frac{d}{t}$$

ROCKET BALLOONS

#1

Topic
Rocket balloons

Key Question
1. How does a rocket work?
2. What gets it off the ground?

Focus
Students will learn how a rocket works.

Math
NCTM Standards
- *Estimate, make, and use measurements to describe and compare phenomena*
- *Systematically collect, organize, and describe data*

Measurement
   length
Averaging

Science
Project 2061 Benchmarks
- *Tables and graphs can show how values of one quantity are related to values of another*
- *Keep records of their investigations and observations and not change records later*

Physical science
   Newton's Third Law of Motion

Processes
Observing
Collecting and recording data
Interpreting data
Controlling variables

Materials
Sausage-type balloons
Soda straws
1-gal. plastic bag
Masking tape
String (monofilament fishing line works best)
Measuring tape or meter stick

Background Information
This activity uses rocket balloons to illustrate Newton's Third Law of Motion: for every action there is an equal but opposite reaction. The backward thrust of the air from the balloon rocket produces the forward motion of the balloon.

In this initial investigation, we are going to consider the relationship between the length of the balloon and the distance which it travels. As air escapes from the balloon, the rocket will travel along the track (string). Logically, the more air in the balloon, the longer its length and, therefore, the farther it will travel.

Management
1. This activity will take one to two class periods, or 45-90 minutes.
2. Try to have as many different lines strung as possible to use as tracks for the balloons. Hang the strings at eye level.
3. This activity works best in small groups.
4. Assign jobs which can be rotated: one person to (a) launch the balloon, (b) record the distance, (c) observe and check.
5. If room is limited, you will have to lessen the size of the balloons.

Procedure
1. Thread string through a straw and attach the ends of the string to a wall or other object. Stretch the string as tightly as possible.
2. Tape the plastic bag to the straw. This will serve as a "pocket" in which to insert the blown up balloon.
3. Blow up a balloon to the desired size, measure its length, and record the data.
4. Insert the balloon into the plastic bag being careful not to release any air.
5. Release the balloon. Observe and record the distance it travels.

Discussion
1. Why does the balloon travel along the string? [thrust: the backward thrust of the released air creates the forward motion of the balloon]
2. What happened as the length of the balloon was increased? Why? [more air was released]
3. What factors could cause a balloon not to go as far as it should? [the balloon becomes stretched out and does not release its air at the same rate]

ROCKET BALLOONS

CHART #1

Balloon length vs. Distance Traveled

Balloon Length	Distance Traveled			Average Distance
	Trial 1	Trial 2	Trial 3	
1.				
2.				
3.				
4.				
5.				

ROCKET BALLOONS

2

Topic
Rocket balloons

Key Question
Does the distance a balloon rocket travels change as the angle of ascent increases?

Focus
Students will explore the ways in which the angle of ascent might affect a rocket.

Math
NCTM Standards
- *Extend their understanding of the concepts of perimeter, area, volume, angle measure, capacity, and weight and mass*
- *Systematically collect, organize, and describe data*

Measurement
   length
   angles
Averaging

Science
Project 2061 Benchmarks
- *Tables and graphs can show how values of one quantity are related to values of another.*
- *Recognize when comparisons might not be fair because some conditions are not kept the same.*

Physical science
   Newton's Third Law of Motion

Processes
Observing
Collecting and recording data
Interpreting data
Controlling variables
Generalizing

Materials
Sausage-type balloons
Soda straws
1-gal. plastic bag
Masking tape
String (monofilament line)
Measuring tape or meter stick
Protractor

Background Information
This activity builds upon *Rocket Balloons #1*. Instead of comparing the length of balloon to the distance traveled, *Rocket Balloons #2* invites students to compare the distances traveled by balloons moving horizontally, at a 45° angle and at a 90° angle. Since the balloon will have to travel up the line at an angle, some of the thrust will be expended in this effort.

Management
1. This activity works best in small groups and will take 45-60 minutes.
2. Try to have as many different lines strung as possible.
3. Assign jobs which can be rotated – a launcher, a recorder and a checker.
4. Hopefully, if students are careful when checking the measurements of the angles when they move the lines, they will also be careful in the physical process of moving the lines.
5. Again, if space is limited, lessen the size of the balloon.

Procedure
1. Thread three strings through three straws. Attach one string horizontally between two walls. Attach one end of the second string to the floor and raise the other end to make a 45° angle and attach it to the wall. With the third string, attach one end to the floor and the other end to the ceiling to make a 90° angle.
2. Tape a plastic bag to each of the straws to serve as "pockets" for the balloons.
3. Have students blow up the balloons (the same size each time), measure and record their lengths.
4. After the balloons are released, have students measure the distance each balloon travels and compare results.
5. After each distance is recorded, move to a string which is at a different angle. Repeat the above procedure for the different angles of string.

Discussion
1. What happens when the line is changed from level flight to a different angle of flight?
2. Does it take more or less power to lift the balloon rocket "off the ground" when the line is at an angle than it does when the line is level? Why?
3. Does it take more or less force to move the balloon rocket the same distance along an upward path as it does to move it along a level path? Explain.
4. How did your distances compare between the flights at 45° and 90°?
5. Make a generalization about your results.

Extension
Students can calculate the speed at which their rocket balloons travel by dividing the distance traveled by the time it took.

ROCKET BALLOONS

CHART # 2

Angle of Flight   vs.   Distance Traveled

Balloon Size	Distance Traveled		
	Horizontal	45°	90°
1.			
2.			
3.			
4.			
5.			

HORIZONTAL

45° ANGLE

90° ANGLE

ROCKET BALLOONS

3

Topic
Rocket balloons

Key Question
How does the size of the opening affect the rocket balloon's flight?

Focus
The students will investigate the effects on the rocket balloon's flight when its opening is changed.

Math
NCTM Standards
- *verify and interpret results with respect to the original problem situation*
- *describe and represent relationships with tables, graphs, and rules*

Measurement
    length
    angles
Geometry
    diameter
    circumference
Averaging

Science
Project 2061 Benchmarks
- *Learning means using what one already knows to make sense out of new experiences or information, not just storing the new information in one's head.*
- *Results of scientific investigations are seldom exactly the same, but if the differences are large, it is important to try to figure out why. One reason for following directions carefully and for keeping records of one's work is to provide information on which might have caused the differences.*

Physical science
    Newton's Third Law of Motion

Processes
Observing
Collecting and recording data
Interpreting data
Generalizing
Controlling variables

Materials
Sausage-type balloons
Drinking straws and cocktail sipping straws
Masking tape
1-gal. plastic bag
Various devices to regulate size of opening in balloons
    (see *Background Information*)
String (monofilament line)
Measuring tape
Protractor

Background Information
    This activity, *Rocket Balloons #3*, introduces a new variable, the size of the opening of the balloon. The greater the opening, the faster the air will escape. One question is whether the increased force of the thrust is great enough to compensate for the shorter length of time that the air is escaping.

Management
1. As before, this activity works best in small groups, with students rotating the jobs. It will require one to two class periods.
2. Different-sized openings can be made by taping short pieces of straws inside the end of the balloons: drinking straws in two different diameters and cocktail sipping straws would give three variations. You might also try bent paper clips if you watch the sharp ends and roll the mouth of the balloon around the shape created. Wine corks which have been sliced thin and had different sized holes made in them also make good end pieces. Search department stores or hardware stores for small, lightweight plastic rings of various sizes.
3. Make sure that students are consistent when they measure the openings; establish that they will measure only the inside diameter of the opening. (If you have a pair of calipers rusting in the closet, this might be a fun time to bring them out.) Mention *diameter* and *circumference*, if you wish.
4. Have students experiment with fitting end pieces into the mouth of a balloon before testing as this takes

some dexterity.

Procedure

1. The students will construct rocket balloons as they did in *Rocket Balloons #2*, but they will experiment with creating different-sized, measurable openings at the mouth of the balloon.
2. When they have established the method of making the openings, have them measure the diameter of five "mouthpieces" of various circumferences and record the information on their activity page.
3. Have students test each of the five "mouthpieces" using a horizontal track and record the results. Make sure that the balloon used is always the same length.
4. After discussing the results obtained on a horizontal flight, ask students what they think will happen on the angled flights.
5. Finally, have students test their own rocket balloons using the same "mouthpieces" but on the 45° and 90° tracks.
6. Record all data and discuss the results.

Discussion

1. What happens if you decrease the size of the opening of the balloon rocket on a horizontal flight? Why?
2. What happens if you decrease the size of the opening of a balloon rocket on an angled flight? Why?
3. How great was the difference in the distances the rockets traveled? What was the average distance traveled by each rocket with the same diameter opening?

Extensions

1. If the students are interested, have them add the variable of different balloon lengths once again. You can challenge them to design the best rocket balloon for each of the three tracks by varying length and the size of the opening.
2. String two parallel lines and run a relay race. Divide the groups into teams. Have them get the first balloon ready. At the word "Go!," each team releases its first rocket. When it stops, the balloon must be re-inflated and released again. When it reaches the opposite wall, it must be turned around and started again. Upon return to the starting place, the second balloon replaces it and covers the course, etc. Students will have to decide for themselves which is the most effective design – opening and length – before the race begins in order to obtain optimum speed and distance in each "heat."

ROCKET BALLOONS

CHART #3

	Balloon Size	Size of Balloon Opening	Distance Traveled		
			Horizontal	45°	90°
1.					
2.					
3.					
4.					
5.					

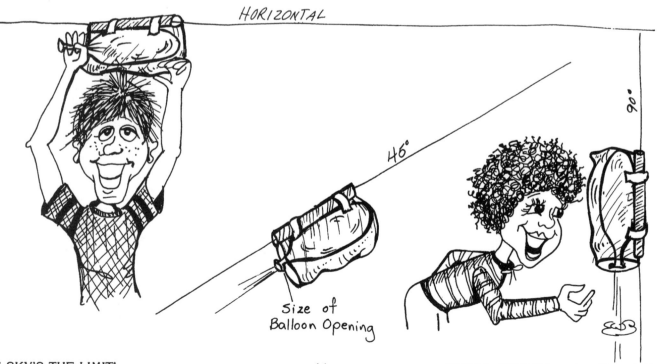

HORIZONTAL

45°

90°

Size of Balloon Opening

Topic
Rocket balloons

Key Question
How can you overcome gravity and make a balloon hover in the air in a stable position?

Focus
Students will learn more about how a rocket works, how the force of the engines can be controlled.

Math
NCTM Standards
- *Model situations using oral, written, concrete, pictorial, graphical, and algebraic methods*
- *Apply mathematical thinking and modeling to solve problems that arise such as art, music, pschology, science, and business*

Measurement
  time
Averaging
Estimating

Science
Project 2061 Benchmarks
- *Even a good design may fail. Sometimes steps can be taken ahead of time to reduce the likelihood of failure, but it cannot be entirely eliminated.*
- *Recognize when comparisons might not be fair because some conditions are not kept the same.*

Processes
Observing
Predicting
Controlling variables
Collecting and recording data
Interpreting data
Applying and generalizing

Materials
Balloons – long and/or round
Drinking straws
Masking tape
Stopwatches
Ruler

Background Information
This investigation will demonstrate to students how a spaceship may someday land on Earth or how the LEM (Lunar Escape Module) piloted by Neil Armstrong first landed on the moon. Certainly students realize that an object landing in a vertical position must slow its speed so as not to impact the landing areas with such great force that severe damage to craft and pilot will result. Here, students will control the escape of air from the balloon in such a way that the force of the thrust will be sufficient only to keep the balloon up, not enough to make it go up.

This is also the principle that is used by the British Royal Air Force in the Harrier Jet, a jet airplane that is able to take off or land in a vertical position and also hover in a stationary position, as a helicopter does.

Management
1. The size and shape of the balloon can cause differences in the way the balloons will fly. At first, give all students balloons of the same size and shape. Later, size and shape may be added as a variable or extension.
2. The size of the straw (weight and diameter) will affect the flight of the balloon. This may also be added later as a variable or extension.
3. Make sure the straws are taped to the balloon securely so that no leaks exist.
4. Groups of three will work best, with one student to release the balloon, one observer to tell when to start and stop the clock, and a timer. A fourth student could be added for recording.
5. Allow 45-60 minutes for construction, flying, recording, and discussion.

Procedure
1. Each group will need at least one balloon, two (or more) drinking straws, masking tape, and a timing device.
2. To adjust the length of the straws, students need to crimp the end of one straw and insert it into the end of the other straw.
3. Insert the straw(s) into the end of the balloon and tape so no leaks occur.
4. With observers at the ready, measure and record the length of the straw, inflate the balloon and release it.
5. The object of this activity will be to adjust the straws in such a way so as to keep the balloon up (in the air) as long as possible.
6. Repeat as desired or necessary.

Discussion
1. What does make an aircraft hover? [The force of the engines is constricted until there is only enough force left to overcome the force of gravity or the weight of the craft.]
2. What was the most difficult part of this activity? Can you think of any ways to make it easier?
3. Find other groups with similar straw lengths and compare their results? If the results are not the same, try to explain them.

HOVER CRAFT

_____ NAME

Record Flight Time in Seconds

STRAW LENGTH	cm	cm	cm	cm	cm	cm	cm
FLIGHT #1							
FLIGHT #2							
FLIGHT #3							
TOTAL FLIGHT TIME							
AVERAGE FLIGHT TIME							

Type of Balloon _____ (Round, Sausage, etc.)

Longest Flight Time _____ (Time Aloft)

Construction:

1. Crimp Straw and Insert into second.

2. Measure Straw Length + Record. Insert Straws Into Balloon and Tape - No Leaks!

3. Inflate and Release. Time the Flight and Record.

What is The Best Length Of Straw For Your Balloon?

THE SKY'S THE LIMIT!

46

©1994 AIMS Education Foundation

WATER ROCKETS

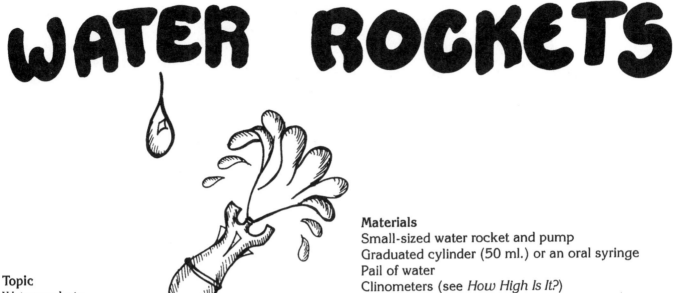

Topic
Water rockets

Key Question
How would varying the amount of pressure and/or the amount of water affect the thrust of a water rocket?

Focus
Through the use of toy water rockets, students will explore the effects of variables on thrust.

Math
NCTM Standards
- *Extend their understanding of the process of measurement*
- *Estimate, make, and use measurements to describe and compare phenomena*
- *Develop the concepts or rates and other derived and indirect measurements*

Measurement
    length
    angles
Using the tangent function

Science
Project 2061 Benchmarks
- *Tables and graphs can show how values of one quantity are related to values of another.*
- *Measurements are always likely to give slightly different numbers, even if what is being measured stays the same.*

Physical science
    thrust
    Newton's Third Law of Motion

Processes
Observing
Collecting and recording data
Interpreting data
Identifying and controlling variables
Predicting
Generalizing

Materials
Small-sized water rocket and pump
Graduated cylinder (50 ml.) or an oral syringe
Pail of water
Clinometers (see *How High Is It?*)
15 meters string

Background Information
We are dealing with the concept of action versus reaction. In aerodynamics, this is called "thrust," which creates lift by overcoming the force of gravity. The thrust can be measured as it correlates directly to the altitude the rocket attains.

The variables investigated in this activity are the amount of water in the rocket and the air pressure, which is increased by pumping. More water makes the rocket heavier; a greater number of pumps creates more air pressure.

The chart to determine the altitude of the rocket is based on trigonometric principles. Although knowledge of these principles is not essential to the activity, a basic explanation follows in case (oh, help!) students ask.

In any right triangle, there are specific relations between the lengths of any two sides. These are determined by the measures of the angles. Considering a, the three sides of a right triangle can be identified as the one "adjacent" to the angle, the one "opposite" the angle, and the "hypotenuse" of the triangle. The three trigonometric functions are sine, cosine, and tangent (the one with which we're concerned). These express ratios between the lengths of the sides of the triangle as follows:

$$\text{sine } a = \frac{\text{OPPOSITE}}{\text{HYPOTENUSE}}$$

$$\text{cosine } a = \frac{\text{ADJACENT}}{\text{HYPOTENUSE}}$$

$$\text{tangent } a = \frac{\text{OPPOSITE}}{\text{ADJACENT}}$$

A useful mnemonic for remembering these (since they are always printed in the same order in the tables) is, "**O**scar **h**ad **a** **h**eap **o**f **a**pples."

OSCAR	A	OF
HAD	HEAP	APPLES

OPPOSITE	ADJACENT	OPPOSITE
HYPOTENUSE	HYPOTENUSE	ADJACENT

sin a	cos a	tan a

As you can see from *Tangent of an Angle*, the tangent is a four-place decimal; if you divide the length of the side opposite an angle (in a right triangle) by the length of the side adjacent to it, that is the number you will obtain. It is unique to the measurement of the angle in the triangle. In this activity, we know the angle and its tangent, and we know the length of the adjacent side; therefore, we could find the altitude by multiplying the tangent by 15 meters. (For clinometer use, please refer back to *How High Is It?*)

Management

1. Time suggested: one 30-minute period for orientation; one 45-minute period outside for shooting water rockets; a 30-minute (or more) period for making charts and graphs and comparing results.
2. To launch the water rocket, follow the directions on its package. Prior practice is recommended.
3. Students should work in groups of three to six and rotate jobs.
   a. Clinometer sighter
   b. Clinometer reader
   c. Data recorder
   d. Observer(s)
4. Tie a pencil or other pointed object to the end of the 15 meter string. Have someone hold the string while you walk around dragging the pointed object in the dirt at the launch site. This will mark a circular area with a radius of 15 meters. Groups should be equidistantly spaced around the circumference.
5. Students should have done the activity *How High Is It?* and be comfortable using clinometers.
6. Be sure to keep the pump of the water rocket well oiled.
7. Practice tracking an object thrown into the air and sighting it at its highest point. This will help improve accuracy of readings.
8. The first activity page contains a table which assumes students will stand 15 meters from the launching mark. The second activity page contains a table in which students can stand at various distances and calculate the height of the rocket.

Procedure

1. In the classroom, form groups and explain jobs and rotational procedures.
2. Collect materials and go outside to an open area.
3. Go to the launch site and position groups around the circle.

4. Use a syringe (or a graduated cylinder and a funnel) to fill the rocket with 10 ml of water. Pump five times.
5. Launch the rocket vertically. (Keep an eye on the rubber gasket as it comes off easily.)
6. Groups will use clinometers to measure the angle of the rocket at its highest point and record that angle on their activity sheet.
7. Repeat this process for different volumes and pumps (as suggested on the activity sheet). Students can rotate jobs.
8. When all data are recorded, return to classroom to do the calculations.

Discussion

1. How is thrust affected by increasing the pressure while the volume of water stays the same? [The thrust is increased.] How do you know?
2. How is thrust affected by increasing the volume of water when the pressure remains the same? [The thrust is increased.]
3. Can you find any ratios relating the altitude to pressure or water volume? [Graphing would help; answer depends on individual results.]
4. What principles of physics have been experienced in this activity? [Newton's Third Law – For every action there is an equal and opposite reaction.]
5. What are some of the problems that need to be overcome in this activity? [Sometimes the water rocket doesn't go straight up.] How could these problems be overcome or compensated for?

Extensions

1. Use a blank activity sheet to try additional combinations of pressure and water. Can you find an optimum volume of water at a given number of pumps for maximum thrust or altitude?
2. Try the same volume of ice water and warm water at the same pressures. Compare the results.
3. Try a different liquid (salt water or rubbing alcohol) and compare results with those of regular water.
4. Use the formula in *How High Can You Throw It?* to calculate the altitudes.
5. Use a rubber tube, a container of water and a graduated cylinder to calculate the volume of air in each pump. Compare this to the volume of the rocket. Any conclusions?

WATER ROCKETS

Sighter

_____

Reader

_____

Recorder

_____

Observer's Point

_____

Height in Meters

70°
65°
60°
55°
50°
46°
40°
35°
30°
25°
20°
15°
10°
5°

40
39
38
37
36
35
34
33
32
31
30
29
28
27
26
25
24
23
22
21
20
19
18
17
16
15
14
13
12
11
10
9
8
7
6
5
4
3
2
1
0

Take Off Point

15 METERS

WATER ROCKETS

NAME _____

TRIAL NUMBER	VOLUME OF WATER	NUMBER OF PUMPS	ANGLE°	CLASS AVERAGE ANGLE°	ALTITUDE
1	10 ml	5			
2	10 ml	10			
3	10 ml	15			
4	10 ml	20			
5	20 ml	5			
6	20 ml	10			
7	20 ml	15			
8	20 ml	20			
9					
10					

THE SKY'S THE LIMIT!

50

WATER ROCKETS

NAME

TRIAL NUMBER	VOLUME OF WATER	NUMBER OF PUMPS	ANGLE	TANGENT OF ANGLE	× DISTANCE FROM ROCKET	= HEIGHT OF ROCKET
1						
2						
3						
4						
5						
6						
7						
8						
9						
10						

PARACHUTE INVESTIGATION

Tangent of Angle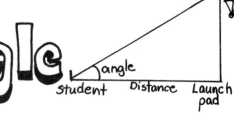

Angle°	Tangent	Angle°	Tangent	Angle°	Tangent
1	.01746	31	.6009	61	1.804
2	.03492	32	.6249	62	1.881
3	.05241	33	.6494	63	1.963
4	.06993	34	.6745	64	2.050
5	.08749	35	.7002	65	2.145
6	.10510	36	.7265	66	2.246
7	.12278	37	.7536	67	2.356
8	.14054	38	.7813	68	2.475
9	.15838	39	.8098	69	2.605
10	.1763	40	.8391	70	2.747
11	.1944	41	.8693	71	2.904
12	.2126	42	.9004	72	3.078
13	.2309	43	.9325	73	3.271
14	.2493	44	.9657	74	3.487
15	.2679	45	1.0000	75	3.732
16	.2867	46	1.0355	76	4.011
17	.3057	47	1.0724	77	4.331
18	.3249	48	1.1106	78	4.705
19	.3443	49	1.1504	79	5.145
20	.3640	50	1.1918	80	5.671
21	.3839	51	1.2349	81	6.314
22	.4040	52	1.2799	82	7.115
23	.4245	53	1.3270	83	8.144
24	.4452	54	1.3764	84	9.514
25	.4663	55	1.4281	85	11.430
26	.4877	56	1.4826	86	14.301
27	.5095	57	1.5399	87	19.081
28	.5317	58	1.6003	88	28.64
29	.5543	59	1.6643	89	57.29
30	.5774	60	1.7321	90	——

Tangent of Angle ✗ Distance from launch = Height

It's The Last Straw!

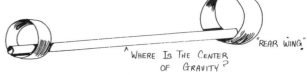

"REAR WING"

^ WHERE IS THE CENTER OF GRAVITY?

Topic
Loop airplanes

Key Question
How far will a loop airplane fly?

Focus
Using a straw and two strips of paper, students will make a paper airplane which really does fly. They will use their planes in recording distance, graphing, and in working the metric system.

Math
NCTM Standards
- *Systematically collect, organize, and describe data*
- *Construct, read, and interpret tables, charts and graphs*
- *Understand the structure and use of systems of measurement*

Measurement
   length
Averaging

Science
Project 2061 Benchmarks
- *The scale chosen for a graph or drawing makes a big difference in how useful it is.*
- *Measurement instruments can be used to gather accurate information for making scientific comparisons of objects and events and for designing and constructing things that will work properly.*

Physical science
   aerodynamics

Processes
Observing
Collecting and recording data
Interpreting data
Controlling variables
Generalizing

Materials
Straws – one per airplane
Transparent tape
Metric measuring tapes
Optional: poster board for a class graph

Background Information
The principle joy of this activity is that there is little background information! Thanks to the principles of lift and thrust, this rather unusual object actually does fly! Enjoy!

In the process of enjoying, have the students practice converting from one metric unit to another as they measure the distances their planes fly. The student sheet asks for measurements in centimeters and meters; students may be encouraged to make complete charts.

Older students can use this experience to use scientific notation. If the plane flew 2.275 m, show them that the distance can be expressed in other units without changing the number:

$$2.275 \text{ m} \quad \begin{aligned} &= 2.275 \times 100 \text{ m} \\ &= 2.275 \times 10^1 \text{ dm} \\ &= 2.275 \times 10^2 \text{ cm} \\ &= 2.275 \times 10^3 \text{ mm} \end{aligned}$$

Management
1. Each student will build a plane according to the pattern, but it will be easier to measure the distance flown if students work in pairs. One student will launch a plane while the other marks the distance it flew.
2. Allow 40-50 minutes.
3. You will want the students to establish guidelines for measuring the length of the flights. They will need to decide whether to measure the distance at which the plane touches the ground or where it stops. They also need to decide whether to measure the front of the plane or the rear.
4. You may want students to do *This is Definitely The Last Straw* first as the pattern presented here may be difficult to improve upon.

Procedure
1. You might introduce the activity by showing the students a straw and two loops of paper asking whether they think a "flying object" could be constructed from them.
2. Have students construct loop airplanes making them as similar to the illustration as possible so comparisons will be easier to make.
3. Have students identify their planes by using the *Aviation Alphabet* included in this book. Direct them to put their identification on the rear wing of the planes.

4. Establish the rules for fair measurement.
5. Have students make five flights for each plane, measure the distance flown in centimeters, and record in centimeters and meters.
6. Direct them to find the average distance flown.
7. To graph their data, students will need to decide on a scale for the *Distance of Flight* axis. Utilize this time to let students discover that the more graphing space they use, the more refined their interpretations can be.

Discussion
1. Where is the center of gravity of your loop plane? How do you find out?
2. How and where do you hold your plane to launch it? Why? What happens if you try to fly it the other way around? Why?
3. Is the fact that the straw is hollow important? Try plugging one end. What happens?

Extension
1. Interested students can make a table showing the distances expressed in all the metric units.
2. Make a class graph showing the maximum flight distance for each student's plane. Find the mean, median, and mode of the distances measured.
3. Make a frequency table showing the distances flown (maximum from each student). Determine the range of the set of numbers. Have each student determine the percent variation from the mean for his plane.
4. Discuss the probability of a plane's falling short of or exceeding the mean distance flown.
5. Use the recorded measurements to work with scientific notation.

It's The Last Straw!

USE A STRAW AND TWO STRIPS OF PAPER TO MAKE A FLYING OBJECT

"REAR WING"

^ WHERE IS THE CENTER OF GRAVITY?

USE THE PATTERNS GIVEN FOR THE STRIPS (A, B, ON NEXT PAGE). CUT OUT AND TAPE TO THE ENDS OF YOUR STRAW AS SHOWN. PRINT YOUR NAME OR THE "PLANES" NAME ON THE REAR WING.

MAKE FIVE TEST FLIGHTS. FOR EACH, MEASURE THE DISTANCE FLOWN IN CENTIMETERS BUT RECORD IN BOTH CENTIMETERS AND METERS.

DISTANCE

FLIGHT No.	DISTANCE FLOWN IN	
	CENTIMETERS	METERS
1		
2		
3		
4		
5		
TOTAL		
AVERAGE		

FLIGHT DISTANCE GRAPH

FLIGHT No.

1

2

3

4

5

DISTANCE OF FLIGHT

PATTERNS

STRAW PLANE BANDS

THESE STRIPS WILL CREATE A STRAW PLANE WITHOUT A STRAW!!

THESE PIECES WILL CREATE ANOTHER VERSION OF THE "STRAW PLANE". BANDS C AND D ARE USED IN THE SAME WAY AS IN THE FIRST VERSION.

A

B

C

D

E

BAND E REPLACES THE STRAW. CUT OUT E AND SCORE ALONG DOTTED LINES. OVERLAP 2 SIDES AND GLUE TO FORM A TRIANGULAR ROD. GLUE THE ROD INSIDE THE RINGS.

This is Definitely
The Last Straw!

Topic
Loop airplanes

Key Question
How can the basic loop airplane be improved?

Focus
Having tested a loop airplane constructed according to a given pattern, students will now experiment in manipulating the variables of design in an effort to improve the basic design.

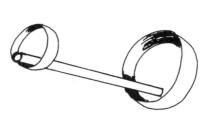

Math
NCTM Standards
- *Systematically collect, organize, and describe data*
- *Model situations using oral, written, concrete, pictorial, graphical, and algebraic methods*

Measurement
  length
Averaging

Science
Project 2061 Benchmarks
- *Measuring instruments can be used to gather accurate information for making scientific comparisons of objects and events and for designing and constructing things that work properly.*
- *If more than one variable changes at the same time in an experiment, the outcome of the experiment may not be clearly attributable to any one of the variables. It may not always be possible to prevent outside variables from influencing the outcome of an investigation (or even to identify all of the variables), but collaboration among investigators can often lead to research designs that are able to deal with such situations.*
- *Know that often different explanations can be given for the same evidence, and it is not always possible to tell which one is correct.*

Physical science
  aerodynamics

Processes
Observing
Controlling variables

Collecting and recording data
Interpreting data
Generalizing

Materials
Drinking straws
Paper – perhaps of different types and weights
Transparent tape
Rulers
Metric tape measures

Background Information
There are three basic variables involved in the design of this airplane: (a) the size of the loops (both width and length, or circumference); (b) the material used to make the loops; and (c) the placement of the loops. Students can also select either the straw or the folded-paper fuselage, but to control this variable, they should select one or the other.

Since the circumference of the circle made by the loop is known, challenge students to determine the radius of each loop used. $c = 2\pi r$, so r (radius) $= c/2\pi$. A good calculator activity that could be added at this time would be to use the length of the loop as the circumference (subtract the overlap used for gluing or taping; use $\pi = 3.14$.

Do not feel restricted to size by the straw length; the folded-paper fuselage allows students to make larger models.

Management

1. Encourage students to proceed somewhat logically. They might begin by predicting what would happen if they shortened the fuselage of the prototype model by moving the loops closer together. Allow ample time for predictions and exploration so they can manipulate one variable at a time.
2. Once each student has developed two models he/she finds satisfactory, test them for distance as before, using partners and allowing five trials.
3. If desired, when all the serious record-keeping has been done, hold an all-out, one-chance championship match to see which student's loop plane flies the longest distance.
4. Allow approximately 40-60 minutes for this activity.

Procedure

1. Discuss the variables involved with the loop airplanes and have students predict the results of certain changes.
2. Each student will construct two different models of the loop plane.
3. Have students record all data regarding each of their models and test each for distance.
4. Direct students in a discussion of their results and attempt to decide what factors of design influenced performance most positively.

Discussion

Before designing loop planes
1. What could be changed in the design of the loop airplane?
2. What effect might each change have on the distance the plane would fly?
3. Can you think of anything you definitely would not want to do to the design of the plane?

After designing and testing new loop planes
1. What changes most improved the plane? Explain.
2. Were your predictions accurate?
3. After listening to others describe their designs and test results, how would you design the ultimate plane?

This is Definitely
The Last Straw!

Now, You Can Modify The Basic Design In Any Way You Wish: Make The Bands Longer, Shorter, Wider, Narrower; Change The Center Of Gravity Or Whatever You Wish.

As You Work, Choose Your 2 Best Models And Record The Descriptive Information Below.

MY ORIGINAL MODELS

SPECIFICATIONS	MODEL #1	MODEL #2
NAME		
LENGTH OF STRAW IN CM		
SHORT BAND LENGTH IN CM		
SHORT BAND WIDTH IN CM		
LONG BAND LENGTH IN CM		
SHORT BAND WIDTH IN CM		
CENTER OF GRAVITY FROM FRONT IN CM		
CENTER OF GRAVITY FROM BACK IN CM		
RECORD DISTANCE FLOWN IN METERS MAKE 5 FLIGHTS OF EACH MODEL	#1	#1
	#2	#2
	#3	#3
	#4	#4
	#5	#5
TOTAL		
AVERAGE		

WHAT ARE SOME MODIFICATIONS THAT MIGHT IMPROVE YOUR DESIGN?

First Class Airplanes

Introduction

You and your students are about to embark on another very satisfying and interesting experience in the study of flight. You are going to do this without sophisticated equipment and materials, and with little or no expense. Your flying models will be made of paper. Students of all ages enjoy the art of building and flying paper airplanes; therefore, we can turn this enjoyment into a real learning experience.

In this section, you have been provided with many different paper airplane designs, student activity sheets (for data collection), directions, management suggestions, lists of materials, and an extension activity - the *Air Fair* - that probably will spill over from your classroom to other parts of the building.

You will start activities by having students select and build the five different paper airplane designs provided in this book or designs of their choice. Students may name their aircrafts using the guidelines given in the *Aviation Alphabet*, just as the actual planes are named. The next step will involve the flying and testing of the five different models, and recording the resulting information on student data sheets. There are four different data collection sheets of flight performances. There is one sheet each for *Distance*, *Duration* or flight time, *Accuracy*, and *Aerobatics*. Students will fly airplanes for each category, record the data, and determine which airplane is best suited for a particular type of flying. The best distance or accuracy flyer may not be the best stunt plane. As students discover these facts, they will come to understand some interesting aspects and advantages of certain elements of airplane design.

As an extension to this activity, you may want to organize the *Air Fair,* a paper airplane competition in these four areas, the rules and directions for which you will find at the end of this section.

Good luck and good flying!

Standard Dart

1. Fold a piece of 8.5" X 11" scratch paper in half lengthwise and open.

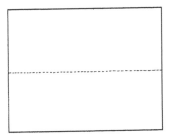

2. Fold the top corners to the center line.

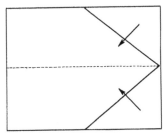
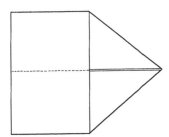

3. Fold the newly formed upper corners to the center line.

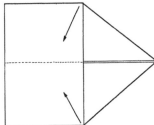
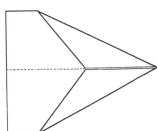

4. Fold in half along the center line and rotate 90°. Mark a point on the left that is one-fourth of the way up from the fold. Draw a line from this point to the tip of the plane. Fold the wing down along this line. Turn over and repeat for the other side.

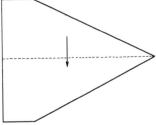
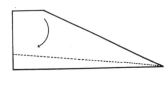

5. Hold the dart underneath and launch with a hard, level throw. Bend the back corners of the wings up or down to adjust the flight.

Roaming Ranger

1. Fold a piece of 8.5" X 11" scratch paper in half lengthwise. Unfold and lay flat.

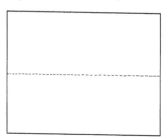

2. Fold the top right corner towards the center fold. Crease the fold. Repeat this step using the top left corner.

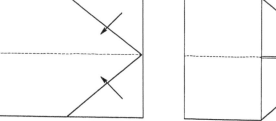

3. Fold down the entire triangle produced in the previous step. Crease the fold.

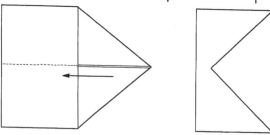

4. Measure 1 1/2" up from the tip of the triangle and make a mark.

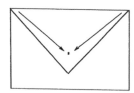

5. Fold in the top corners so that the tip of each corner touches the mark you just made. Crease the folds.

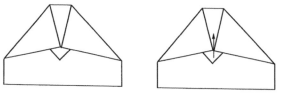

6. Fold up the little triangle so that it covers the corners of the folds made in Step 6.

7. Turn the paper over and fold up along the center line.

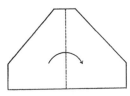

8. Create the wings by folding each side down so that the wing's outer edge is flush with the bottom of the plane. Crease the folds. Voilá! You have a paper plane.

  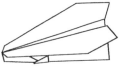

Note: Try making variations of the basic design by altering the folds in the last step. Stabilizers and ailerons can be devised by various cuts and folds on the tail end of the wings.

Ayre Rider

1. Position a piece of 8.5" X 11" scratch paper in front of you so that the width of the paper is facing you. Fold the top right corner down so that the upper edge of the paper is flush with the left edge of the paper. Crease the fold and then reopen the paper. Repeat this step using the left to corner, making it flush with the right edge of the paper.

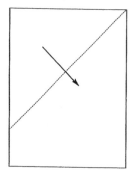 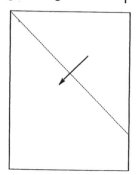

2. You should now have two creases in the paper which cross each other and delineate four sections. Draw the left and right sections in towards the center of the paper. Draw the top section down on top of the left and right sections and press down. Make certain that the outer edges of the top section match evenly with the edges of the bottom section.

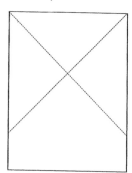 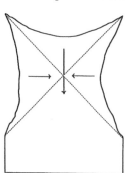 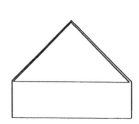

3. Fold about 1" of the tip of the triangle down towards you.

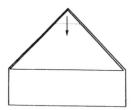 

4. Fold each of the floating sections so that their tips are tucked into the fold made in Step 3. Crease the folds.

5. Fold the sections created in *Step 4* upwards and tuck their uppermost corners into the triangular pocket made in *Step 3*.

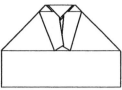

6. Turn the glider over and position it so that its nose is facing to the side. Next, fold the wing tips upwards about 1/2" on each side.

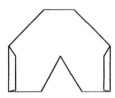

7. Finally, find the center of the wing span and cut equal right triangles on both sides of that point. Make the cuts so that they penetrate the wing just up to the point where the fuselage begins.

Note: Variations can be made on this design by changing wing formations, wing tilt, and by the addition of ailerons and different-sized stabilizers. You can also omit *Step 7* if you wish and leave the wing whole. You may want to experiment with different-sized weights in the nose of the craft.

The *Ayre Ryder* is a glider, not an airplane. As such it requires only a gentle push/drop to fly. The *Ayre Ryder* is ideal for "rooftop" flying.

Flying Wing

1. Fold a piece of 8.5" X 11" scratch paper in half lengthwise and open.

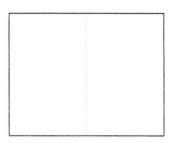

2. Fold the bottom edge to the middle crease.

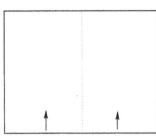

3. Fold again to the middle, making four thicknesses.

4. Flip the four thicknesses over the middle crease so that the middle crease becomes the lower edge. Tape in three places.

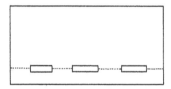

5. Crease the folded section at its midpoint to create a slight angle in the wing.

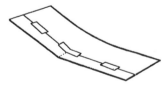

6. Hold the front of the wing at the crease with the index finger on top and the thumb underneath. Launch gently. Bend the back corners up or down slightly to adjust the flight.

Graceful Glider

1. Fold a piece of 8.5" X 11" scratch paper in half lengthwise and widthwise and open.

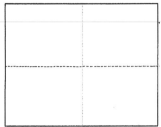

2. Place the paper in front of you so that the length of the paper is facing you. Fold the bottom edge to the middle crease. Fold again to the middle, making four thicknesses.

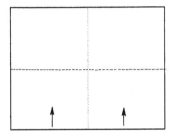
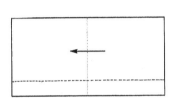

3. Flip the four thicknesses over the middle crease so that the middle crease becomes the lower edge. Tape in three places.

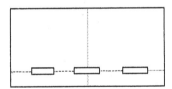

4. Fold in half along the vertical crease. Cut out as illustrated.

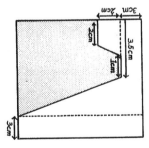

5. Spread the wings out, fold down both sides of the tail, and launch by holding the plane high over your head and releasing with a very gentle thrust.

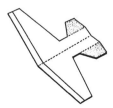

Distance Data

First Class Airlines

Topic
Paper airplanes

Key Question
How does the design affect the flight of a paper airplane in terms of the distance it flies?

Focus
Students will learn the effects of design on the distance a plane can fly.

Math
NCTM Standards
- *Estimate, make, and use measurements to describe and compare phenomena*
- *Acquire confidence in using mathematics meaningfully*

Geometry
Measurement
    length
Estimation

Science
Project 2061 Benchmarks
- *Learning means using what one already knows to make sense out of new experiences or information, not just storing the new information in one's head.*
- *When people care about what is being counted or measured, it is important for them to say what the units are (three degrees Fahrenheit is different from three centimeters, three miles from three miles per hour).*

Physical science
    aerodynamics

Processes
Observing
Predicting
Collecting and recording data
Interpreting data
Generalizing

Materials
Scrap paper - unwrinkled, mostly 8 1/2" x 11"
Metric measuring tapes
Optional: stake or nail (see *Management*)

Management
1. The teacher should pick a fairly calm day, or use a large indoor space for testing. This can be used as an indoor or outdoor activity; whichever you choose to do, make sure you have planned for a starting line, an efficient way to measure distance, and a method for students to retrieve their planes.
2. Each student should build five types of airplanes, either selected from thsi e=section of the book or those of his/her own design.
3. Have plenty of paper handy so that students can build and refine several designs. Because the weight of the paper adds an interesting variable, provide, if possible, different weights.
4. Once the students have built their planes, you may want to do each of the four investigations involved in this section of the book on separate days. Testing itself should not take more than 30 minutes unless you have a very large class.
5. After individuals have tested their different models, you may want to divide the class into small groups to compare their best versions. Group results can be recorded on *Group Distance Competition*.
6. Prior to the activity, establish a starting line. Use masking tape if indoors, or a string tied to stakes if outside.
7. Optional: Two other strings that are pre-marked in meters can be tied to the stakes and used to measure flight distance.

Procedure

1. Have students build five different paper airplanes.
2. Have the students approach the starting line and try to fly the airplanes as far down range as possible.
3. The student will measure the distance from the starting line to the landed airplane and record the measurement on the student data sheet.
4. Repeat as necessary for all five airplanes until the data sheet is completed.
5. Have students compete in small groups using the model which was most successful of their five.
6. Optional: Have students compare the performance of their most successful plane with those of others in a small group.

Discussion

Suggest that students think about and discuss the special features of the plane which flew the longest distance.

Distance Data

_____ PILOT

PREDICT WHICH OF YOUR PLANES WILL FLY THE FARTHEST.
LIST YOUR PREDICTIONS IN ORDER

PREDICTION	PLANE'S NAME	DISTANCE			TOTAL	AVERAGE
		TRIAL 1	TRIAL 2	TRIAL 3		
1st						
2nd						
3rd						
4th						
5th						

MY BEST PLANE FOR DISTANCE IS_____ .

Group Distance Competition

PICK YOUR BEST PLANE FOR DISTANCE AND COMPETE
WITH YOUR AIRWING.

	PILOT	PLANE	DISTANCE			TOTAL	AVERAGE
			TRIAL 1	TRIAL 2	TRIAL 3		
1.							
2.							
3.							
4.							
5.							
6.							

(BEST AVERAGE DISTANCE)

PILOT   PLANE   DISTANCE

ACES

(BEST DISTANCE)

PILOT   PLANE   DISTANCE

Length Aloft

First Class Airlines

Topic
Paper airplanes

Key Question
How does the design affect the flight of a paper airplane in terms of the length of time it will stay in the air?

Focus
Students will learn the effects of design on how long a paper airpalne will stay in the air.

Math
NCTM Standards
  • *Systematically collect, organize, and describe data*
  • *Acquire confidence in using mathematics meaningfully*
Geometry
Measurement
    time
Estimation

Science
Project 2061 Benchmarks
  • *Measuring instruments can be used to gather accurate information for making scientific comparisons of objects and events and for designing and constructing things that will work properly.*
Physical science
    aerodynamics

Processes
Observing
Predicting
Collecting and recording data
Interpreting data
Applying and generalizing

Materials
Paper planes from *Distance Data*
Stopwatches

Management
1. Follow the suggestions for *Distance Data*.
2. It is best to have students work together on the activity. While one flies the airplane, the other can be the timer. After each trial, the students may change positions as timer and pilot.
3. If your students are doing the comparisons in small groups, let each person test his/her planes individually, sharing the stopwatch with the group; then, each can select the best of the five and compete.

Procedure
1. The students will test the five paper airplanes they have constructed to find out which one will stay in the air for the longest time.
2. The "pilots" will launch their airplanes at a signal from the timers.
3. The timers will start the stopwatch the instant the plane is launched, and stop it when the plane hits the ground.
4. Both the timers and the pilots should read the time of the flight and record it on the pilot's data sheet. (If a watch with a second hand is used, the timer will have to read the time as accurately as possible.)
5. Continue the process until all five airplanes have been tried and the data sheet is complete.
6. When the data charts are completed, students should switch roles and repeat the process.

Discussion
1. What features characterized the planes with the longest times aloft?
2. What features were different from those of the planes which flew the longest distances?

Length Aloft
(FLIGHT TIME)

PREDICT WHICH OF YOUR PLANES WILL STAY ALOFT THE LONGEST.
LIST YOUR PREDICTIONS IN ORDER.

PREDICTION	PLANE	TIME ALOFT IN SECONDS				
		TRIAL 1	TRIAL 2	TRIAL 3	TOTAL	AVERAGE
1st						
2nd						
3rd						
4th						
5th						

MY BEST PLANE FOR TIME ALOFT IS _____.

Length Aloft Competition

PICK YOUR BEST PLANE FOR LENGTH ALOFT AND COMPETE WITH YOUR AIRWING.

PILOT	PLANE	TIME ALOFT IN SECONDS				
		TRIAL 1	TRIAL 2	TRIAL 3	TOTAL	AVERAGE

BEST AVERAGE

PILOT   PLANE   TIME

ACES

BEST LENGTH ALOFT

PILOT   PLANE   TIME

On Your Mark - Accuracy

Topic
Paper airplanes

Key Question
In terms of the accuracy of its flight, how does the design affect the flight of a paper airplane?

Focus
Students will manipulate design features of paper airplanes to improve accuracy.

Math
NCTM Standards
- *Estimate, make, and use measurements to describe and compare phenomena*
- *Construct, read, and interpret tables, charts, and graphs*

Measurement
   length
Estimation

Science
Project 2061 Benchmarks
- *Learning means using what one already knows to make sense out of new experiences or information, not just storing the new information in one's head.*
- *Even a good design may fail. Sometimes steps can be taken ahead of time to reduce the likelihood of failure, but it cannot be entirely eliminated.*

Physical science
   aerodynamics

Processes
Observing
Predicting
Collecting and recording data
Interpreting data
Applying and generalizing

Materials
Paper planes used in *Distance Data*
Metric measuring tape
Target (see *Management*)

Management
1. Follow the management suggestions for the previous investigations in this section.
2. You will have to have a starting line and a target for this activity. Targets from *Be A Rotor Promoter* can be used for this activity. Other items that work as targets are softball bases, pieces of butcher paper which are weighted down, or a stake.
3. Have the course laid out before you begin. If you have a large class or limited time, you might set up multiple courses. Some students who finished early in the *Length Aloft* activity might be responsible for setting this one up for the next session.
4. The marked string can be used instead of metric tapes. Tie the string to the target stake. Measurements can be taken in all directions from the target.

Procedure
1. Have each pilot will approach the starting point or line and launch an airplane, trying to get it as close to the designated target as possible.
2. Direct the partner to measure the distance from the target to the landed airplane. The pilot will record the distance on the data sheet.
3. The above steps should be completed with all five airplanes until the record is complete.
4. Again, you may want to have the students compare the flight of their most accurate models by forming small groups.

Discussion
1. Was your most accurate plane different from the ones that were most successful in the other categories?
2. In what ways was it different?
3. What things about your plane do you think affected its accuracy?

On Your Mark - Accuracy

_____ PILOT

PREDICT WHICH OF YOUR PLANES WILL FLY THE MOST ACCURATELY. LIST YOUR PREDICTIONS IN ORDER.

PREDICTION	PLANE NAME	DISTANCE FROM TARGET CENTER				
		TRIAL 1	TRIAL 2	TRIAL 3	TOTAL	AVERAGE
1st						
2nd						
3rd						
4th						
5th						

MY BEST PLANE FOR ACCURACY IS _____

Group Accuracy Competition

PICK YOUR BEST PLANE FOR ACCURACY AND COMPETE WITH YOUR AIRWING.

PILOT	PLANE	DISTANCE FROM TARGET CENTER				
		TRIAL 1	TRIAL 2	TRIAL 3	TOTAL	AVERAGE

(BEST AVERAGE)

PILOT	PLANE	DISTANCE OFF TARGET

ACES

(MOST ACCURATE FLIGHT)

PILOT	PLANE	DISTANCE OFF TARGET

Train Your Plane
Aerobatics

Topic
Paper airplanes

Key Questions
How does design affect the stunts a paper airplane will do?

Focus
Students will manipulate paper airplanes to enable them to perform certain stunts.

Math
NCTM Standards
- *Systematically collect, organize, and describe data*
- *Estimate, make, and use measurements to describe and compare phenomena*

Measurement
  length

Science
Project 2061 Benchmarks
- *There is no perfect design. Designs that are best in one respect (safety or each of use, for example) may be inferior in other ways (cost or appearance). Usually some features must be sacrificed to get others. How such trade-offs are received depends upon which features are emphasized and which are down-played.*

Physical science
  aerodynamics

Processes
Observing
Predicting
Collecting and recording data
Interpreting data
Applying and generalizing

Materials
Paper airplanes from *Distance Data*

Management
1. Keep doing what you have been doing.
2. Give the students time to practice after you have agreed upon the nature of each stunt.
3. Divide the class into small groups and assign each group an area in which to work.

Procedure
Rule: All Planes Must be Launched Horizontally
1. Have students try to make each of their five planes perform the assigned stunts indicated on the student activity sheet. When each stunt has been attempted by a particular plane, the pilot will mark the appropriate column of his record.
2. Continue to proceed through the various stunts until all spaces have either a positive or negative response.

Discussion
1. What were you looking for when you selected the plane you were going to use to do the stunts? How does your answer compare to the criteria of another student in the class?
2. Looking at the planes, which were able to do the most stunts? What sets them apart from the other designs?
3. Were the most successful planes in this category successful in other categories? If so, what categories? If not, why not?

Extensions
1. Have the students draw illustrations on paper to show what is involved in each stunt.
2. Have them write step-by-step instructions for performing the stunts.
3. Have them write a story about "training" their planes.

_____ PILOT

Train Your Plane
Aerobatics

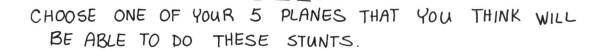

CHOOSE ONE OF YOUR 5 PLANES THAT YOU THINK WILL BE ABLE TO DO THESE STUNTS.

CAN YOU MAKE YOUR PLANE . . . . . .

	YES	NO
FLY STRAIGHT	___	___
CLIMB	___	___
DIVE	___	___
BANK LEFT	___	___
BANK RIGHT	___	___
LOOP	___	___
DOUBLE LOOP	___	___
BOOMERANG	___	___

☆ RULE: ALL PLANES MUST BE LAUNCHED HORIZONTALLY. ☆

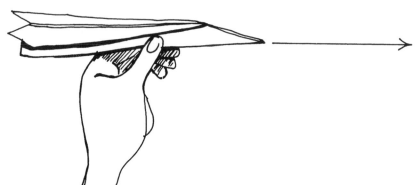

WHICH OF THESE ACROBATIC FEATS DID YOUR PLANE DO BEST?

_____

WHAT DID YOU DO TO YOUR PLANE IN ORDER TO ACHIEVE THIS?

_____

_____

SUPER TUBES

This is a BONUS activity — no tables, graphs or charts. Just make and fly!

There are many unique and amazing designs of paper aircraft. The *Super Tube* offers one of the most amazing flying shapes to come along in quite awhile. If students are asked to guess what is before them, it is seldom that they think this is an aircraft, but it's such fun to fly, they will be easy to convince.

The steps to construct this aircraft are not difficult, but they should be followed closely to assure success. For clarification, see the illustrated page that follows.

Use 8 1/2 " x 11 " paper

Step 1.   Fold corner over and cut off the bottom of the paper to form a square.
Step 2.   Measure or crease along the folded edge to find the center point.
Step 3.   Open flat and fold one corner of the paper to the center point, as shown.
Step 4.   Continue folding to the center (in 1 cm folds) until the fold width meets the center line or diagonal (*Figures 4* and *5*).
Step 5.   Fold once more past the center line (*Figure 6*).
Step 6.   This is one of the special steps that insures success! With the fold up, run the paper over the edge of the table several times to establish a curve. (*Figure 7*), then tape, staple, or glue the overlapped ends.
Step 7.   Hold the *Super Tube* as shown in *Figure 8*, and launch with a gentle!!! toss away from you.

ENJOY! The best things in life ARE free!

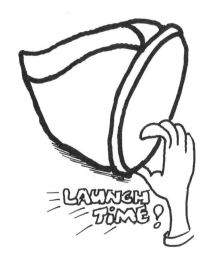

SUPER TUBES

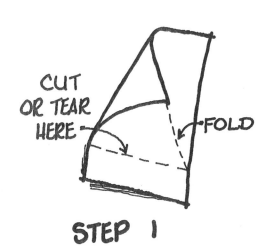

CUT OR TEAR HERE

FOLD

STEP 1

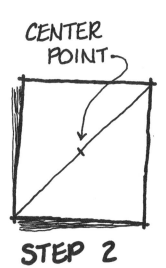

CENTER POINT

STEP 2

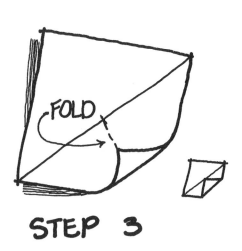

FOLD

STEP 3

KEEP ON FOLDIN'

UNTIL...

STEP 4

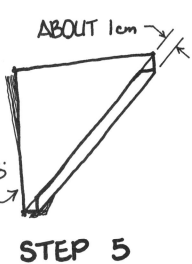

ABOUT 1cm

STEP 5

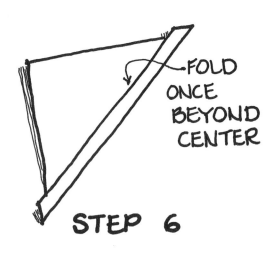

FOLD ONCE BEYOND CENTER

STEP 6

OVERLAP & TAPE TOGETHER

STEP 7

`YE OLDE **FINISHED PRODUCT!!**

LAUNCH TIME!

AVIATION ALPHABET

A	Alpha		N	November
B	Bravo		O	Oscar
C	Charlie		P	Papa
D	Delta		Q	Quebec
E	Echo		R	Romeo
F	Foxtrot		S	Sierra
G	Golf		T	Tango
H	Hotel		U	Uniform
I	India		V	Victor
J	Juliet		W	Whiskey
K	Kilo		X	X-Ray
L	Lima		Y	Yankee
M	Mike		Z	Zulu

Name your plane. Start your identification with N (which stands for United States), then numbers and letters not to exceed a total of 7.

Example ⟨N 952 - SM⟩
November 952 Sierra Mike

⟨N 231 BC⟩
November 231 Bravo Charlie

AIR FAIR

Overview, Organization, and Rules

The *Air Fair* will be broken down into four categories: *Distance*, *Duration* (flight time), *Accuracy*, and *Aerobatics*. Each entrant or pilot will be awarded three points for each event entered during the competition. Awards will be made in each category in addition to one grand "prize" for the overall winner who is the pilot with the largest number of total points.

The *Air Fair* should be organized in the manner described below.

- The contestants will be divided into four fairly equal groups and assigned the numbers one through four. Group One will proceed to the first event (*Distance*), Group Two to the second event (*Duration*), Group Three to the third event (*Accuracy*), and the Group Four to the last event (*Aerobatics*). At the completion of each event, the groups will rotate in numerical order, to the next event: Group One to event two, Group Two to event three, Group Three to event four, and Group Four to event one.
- Each pilot will be provided with an *Entry Card* which will be presented to the judge in charge of the event. The judge will fill in and initial the pilot's card after he/she has competed.
- At the end of the entire competition, each pilot will hand in his/her official Scorecard to the Meet Director who will determine the winners for each individual event and the overall winner.

Organization

Plenty of help will be needed in order to make the meet run smoothly. Suggestions are made in the organization of each event for the number of officials, helpers, and the necessary materials required to run the event properly.

Event #1: Distance

Object: To get the longest flight possible
Personnel: Judge, assistant (optional), two measurers.

Materials: Tape measure, meter stick, or string marked in units of measure.

Procedure: With a starting line established, fix the tape or string with a nail, pencil or any sharp object and stretch it down range. Each pilot will step to the starting line and hand his/her *Entry Card* to the judge. When told to do so, the pilot will launch his/her airplane down range as far as possible. The distance will be marked and measured by the measurers and reported to the judge who will record and initial the appropriate spot on the scorecard. Each pilot will have only one try.

Event #2: Duration

Object: To have the airplane stay in the air as long as possible.
Personnel: Judge, assistant (optional), spotter and timer
Materials: Stopwatch
Procedure: The pilot will approach the judge when it is her/his turn, hand the judge the *Entry Card* and wait until told to launch the airplane. The spotter will yell "go" at which time the pilot will launch his/her airplane and the timer will start timing. When the airplane touches ground, the spotter will yell "stop" at which time the timer will stop the clock. Time (in seconds) is reported to the judge who then records and initials the *Entry Card* and returns it to the pilot. Each pilot will receive only one try.

Event #3: Accuracy

Object: To get as close to a designated point as possible.

Personnel: Judge, assistant (optional), two measurers.

Materials: Target object such as a stake, a measuring device such as a cord marked in units of measure. The marked cord is fastened to the stake so that the distance from the stake (target) to the airplane may be measured in a full 360° circle.

Procedure: Each pilot will hand the judge his/her *Entry Card* and step to the designated starting line. When the pilot is ready, he/she will launch toward the center post trying to land as close to it as possible. The measurers will then rotate the string in a direct line with the airplane and measure the distance from the center post to the airplane and report the distance to the judge who will record and initial the *Entry Card*.

Event #4: Aerobatics

Object: To do as many of the assigned stunts as possible.

Personnel: Judge, helper (optional).
SUGGESTION: Depending on the size of your class, you may want to increase the number of officials to speed up the process at this station.

Materials: None

Procedure: One point will be earned each time an airplane does the assigned stunt. The stunts are in the following order on the *Entry Card:*

1. Straight Flight
2. Climb
3. Dive
4. Right Bank
5. Left Bank
6. Loop
7. Double Loop
8. Boomerang

After each try, the pilot will retrieve his/her craft (or it will be retrieved by the helper) and make ready for the next stunt by making whatever adjustments to the airplane the pilot feels may be necessary. The procedure will continue until all stunts have been attempted. The judge will then score the *Entry Card* and initial it. Each pilot will get one try for each stunt.

Air Fair Rules

1. Each entrant may enter only one airplane for the entire contest; however, the airplane's design may be modified for each category in order to make it perform at the optimum capability.

2. In each event, the pilot will automatically receive three points for entering.

3. Entrants can earn up to 10 points for winning, plus the three additional points for entering, thus giving a maximum point total of 13. The exception will be *Aerobatics* in which one point will be awarded for each stunt the pilot's craft satisfactorily performs.

4. Points will be distributed in the following manner:

   1st place   — 10 points
   2nd place   — 9 points
   3rd place   — 8 points
   4th place   — 7 points
   5th place   — 6 points
   6th place   — 5 points
   7th place   — 4 points
   8th place   — 3 points
   9th place   — 2 points
   10th place — 1 point

5. During the *Aerobatics* competition, the pilot must launch his/her airplane from a level position (about shoulder height), and will receive one point for each stunt completed.

6. If the pilot fails to obtain the signature (initials) of the judge for any event, the pilot will be disqualified for that event only.

7. Each pilot will receive only one try in each event and one try at each stunt during the contest, unless the judge feels it necessary for another try to be granted.

EVENT # 1 DISTANCE

ENTRY CARD ➤

PILOT _____

PLANE _____

AIR FAIR

_____
DISTANCE OF FLIGHT

PLACE FINISH _____
POINTS _____

EVENT # 2 FLIGHT TIME

ENTRY CARD ➤

PILOT _____

PLANE _____

AIR FAIR

_____
FLIGHT TIME IN SECONDS

PLACE FINISH _____
POINTS _____

EVENT # 3 ACCURACY

ENTRY CARD ➤

PILOT _____

PLANE _____

AIR FAIR

_____
MEASUREMENT

PLACE FINISH _____
POINTS _____

EVENT # 4 AEROBATICS

ENTRY CARD ➤

PILOT _____

PLANE _____

FLY STRAIGHT _____
CLIMB _____
DIVE _____
BANK RIGHT _____
BANK LEFT _____
ROLL _____

LOOP _____
DOUBLE LOOP _____
BOOMERANG _____

AIR FAIR

POINTS _____

AIR FAIR ━━▶ MASTER SCORE SHEETS

PILOT	EVENT 1 DISTANCE	EVENT 2 FLIGHT TIME	EVENT 3 ACCURACY	EVENT 4 AEROBATICS	SCORE
1.					
2.					
3.					
4.					
5.					
6.					
7.					
8.					
9.					
10.					
11.					
12.					
13.					
14.					
15.					
16.					
17.					
18.					
19.					
20.					
21.					
22.					
23.					
24.					
26.					
26.					
27.					
28.					
29.					
30.					
31.					
32.					
33.					
34.					
35.					
36.					

THE SKY'S THE LIMIT!

The Whole Kite and Kaboodle

Topic
Construction of paper kites

Key Question
How high can paper kites fly?

Focus
Students will explore the flight characteristics of three simple paper kites.

Math
NCTM Standards
- *Explore problems and describe results using graphical, numerical, physical, algebraic and verbal mathematical models or representations*
- *Extend their understanding of the process of measurement*
- *Extend their understanding of the concepts of perimeter, area, volume, angle measure, capacity, and weight and mass*

Measuring
Charting
Averaging
Graphing

Science
Project 2061 Benchmarks
- *Recognize when comparisons might not be fair because some conditions are not kept the same.*
- *When people care about what is being counted or measured, it is important for them to say what the units are (three degrees Fahrenheit is different from three centimeters, three miles from three miles per hour).*

Processes
Observing
Predicting
Interpreting data
Generalizing

Materials
Paper
Glue
Tape
Hole punch
Kite string or spools of thread
Drinking straws
Crepe paper streamers

Background Information
Students almost invariably ask, "How high is the kite flying?" Accurate measurement of a kite's altitude is not easy. A practical solution is possible, however, if you know just two things: how far away the kite is and the angle between the horizon and a line drawn from you to the kite. That line from you to the kite is the kite line. If all the line is out, you probably know its length.

With the kite's distance known and the angle of its position, you can refer to *Chart A* (meters) or *B* (feet) to determine the altitude. "Belly" or slack in the kite line somewhat affects the accuracy of your calculation at low altitudes. The higher the kite and the steeper the angle, the more nearly equal are the distance and the length of the line.

The kite-flier's question is really very simple: "Is there enough wind to fly my kite?" Take your kite out and hold it up by the bridle, facing the wind. If it assumes a flying posture, you're in business.

Some kites fly in almost a dead calm; others survive a gale. Most flying is done, however, with winds in the range from 4 to 18 miles per hour – and the steadier the better. It is useful, though not necessary, to know how fast the wind is moving.

In 1805, Rear Admiral Sir Francis Beaufort devised his classic wind scale for sailors, rating 0 as dead calm and 12, the top number, as a hurricane. The numbers from 1 to 6, which are germane to kite flying, can be translated like this:
1. 1 - 3 mph   Smoke drifts lazily
2. 4 - 7 mph   Tree leaves rustle
3. 8 - 12 mph   Small flags fly, leaves dance
4. 13 - 18 mph   Tree toss, dust flies, paper flutters
5. 19 - 24 mph   Trees sway, kite strings break
6. 25 - 31 mph   Flying is risky

You can buy simple, dependable wind meters for a few dollars which eliminate guesswork for kite flying as well as expensive elaborate outfits that require fixed installation.

The ground wind and upper wind often differ in velocity, steadiness, and even in direction. But, if you can coax your kite up through the tricky ground wind, there comes that magic moment when the upper wind takes hold, and your kite soars. If you are in the open, the wind's directional variation, usually not more than a few degrees, offers no problem.

Kite Flying Safety Rules
1. Do not fly in the rain.
2. Observe all rules established by local authorities (police, fire, recreation departments, etc.).
3. Do not fly across roads.
4. Do not fly near or across power lines.
5. Do not attempt to retrieve kites lodged in trees, roofs, or power lines.
6. Do not fly in crowded areas (beaches, parks, etc.).
7. Do not fly too close to airports.
8. Do not fly kites that have metal in their constructions or tails.
9. Respect private property at all times.

Management
1. Estimated construction times:
   a. *Paperfold Kite* – 30 minutes
   b. *Dutch Kite* – 45 to 60 minutes
   c. *Straw Kite* – 45 to 60 minutes
2. The classroom teacher is the overall person in charge for the entire activity.
3. Students may be divided into small groups (4,5, or 6 per group).
4. Each group will share in the construction, flying, and recording of information.
5. A flight manual has been provided so students can record their experiences in constructing and flying one of the kites.
6. For kite construction, see individual instructions for *Paperfold Kite, Dutch Kite,* and *Straw Kite.*
7. To use the *Altitude Determining Card* (*ADC*) to determine the altitude of a flying kite:
   a. Each student should have an *ADC*. These cards can be reproduced on tagboard, card stock, or simply glued to cardboard.
   b. The angle and altitude can be determined by using 25 meter or 100 feet kite flying line. Be certain to use whichever scale is appropriate to the length of kite string.(For other lengths of flying lines, use *Chart A* or *Chart B* for determining altitudes.)
   c. Cut out the appropriate *ADC.*
   d. Place the *ADC* on a small table which has been placed on a level surface.
   e. With the end of the kite string dangling over the edge of the table (15 cm or 6 inches), sight the kite using a drinking straw as shown in the illustration on the students' page.

Procedure
1. Have students will construct the kites.
2. The students will record the following information on their *Flight Data Sheet:*
   a. Date of flight
   b. Wind speed (wind meter, call local airport, or refer to Beaufort scale)
3. Allow students to fly their kites.
4. Have them measure the altitude of their kites using different lengths of line.
5. Direct them to record and graph the altitude information.
6. Ask students to find the average height for three or more flights.

Discussion
1. Will your kite fly higher if you use more (longer) string? Explain.
2. What is the highest altitude your kite will fly using any length of string?
3. How much higher will your kite fly in a stronger breeze? Explain your answer.
4. What is the minimum wind speed at which you can fly a kite? How do you know?
5. Can you fly your kite directly overhead? Explain.
6. In the same breeze, will a larger kite fly higher than a smaller kite? Why or why not?
7. Is there a constant ratio between the altitude and the length of the string used? How do you know?

Extensions
1. Make a class graph showing the altitudes of all kites flown with the same length of string. Find the average altitude.
2. Compare the altitudes of kites of the same size that have been constructed from different materials (ditto paper, tagboard, poster board, etc.).
3. Select the length of lines to be used. Have a contest. Which kite flies highest? Which remains aloft longest?
4. Construct kites of different sizes (scaled-up or scaled-down versions of the original kites).
5. Determine the surface area of your kite.

CHART A

Ranged Distance to Kite, or Length of Line (meters)

ALTITUDE OF KITE (METERS)

ANGLE (DEGREES) TO KITE

	10m	15m	20m	25m	30m	35m	40m	45m	50m	75m	100m
5°	.8	1.3	1.7	2.2	2.6	3	3.4	3.9	4.3	6.4	8.6
10°	1.7	2.6	3.4	4.3	5.2	6	6.8	7.8	8.7	12.8	17.4
15°	2.5	3.9	5	6.4	7.8	9	10	11.6	12.9	19.3	25
20°	3.5	5.1	7	8.5	10.3	12	14	15.3	17	25.6	34
25°	4.2	6.3	8.4	10.5	12.6	14.7	16.8	19	21.1	31.6	42.2
30°	5	7.5	10	12.5	15	17.5	20	22.5	25	37.5	50
35°	5.7	8.4	11.4	14.2	16.9	19.9	22.8	25.6	28.6	42.8	57
40°	6.4	9.5	12.8	15.9	19.0	22.3	25.6	28.5	32	48	64
45°	7	10.6	14	17.6	21.2	24.7	28	31.8	35	52.8	70.7
50°	7.7	11.5	14.5	19.1	23	26.8	29	34.5	38.3	57.4	76.6
55°	8.1	12.3	16.2	20.4	24.6	28.6	32.4	37.8	40.6	61.2	81
60°	8.6	12.9	17.2	21.6	25.9	30.3	34.8	38.9	43	64.8	86
65°	9	13.6	18	22.6	27.3	31.7	36	40.8	45.1	67.8	90.6
70°	9.3	14	18.7	23.4	27.9	32.8	37.4	42.2	47	70.4	94
75°	9.6	14.4	19.2	24.1	28.8	33.8	38.5	43.4	48.2	72.4	96.5
80°	9.8	14.7	19.6	24.6	29.5	34.4	39.3	44.2	49.2	73.8	98
85°	9.9	14.9	19.8	24.7	29.7	34.7	39.6	44.6	49.5	74.4	99
90°	10	15	20	25	30	35	40	45	50	75	100

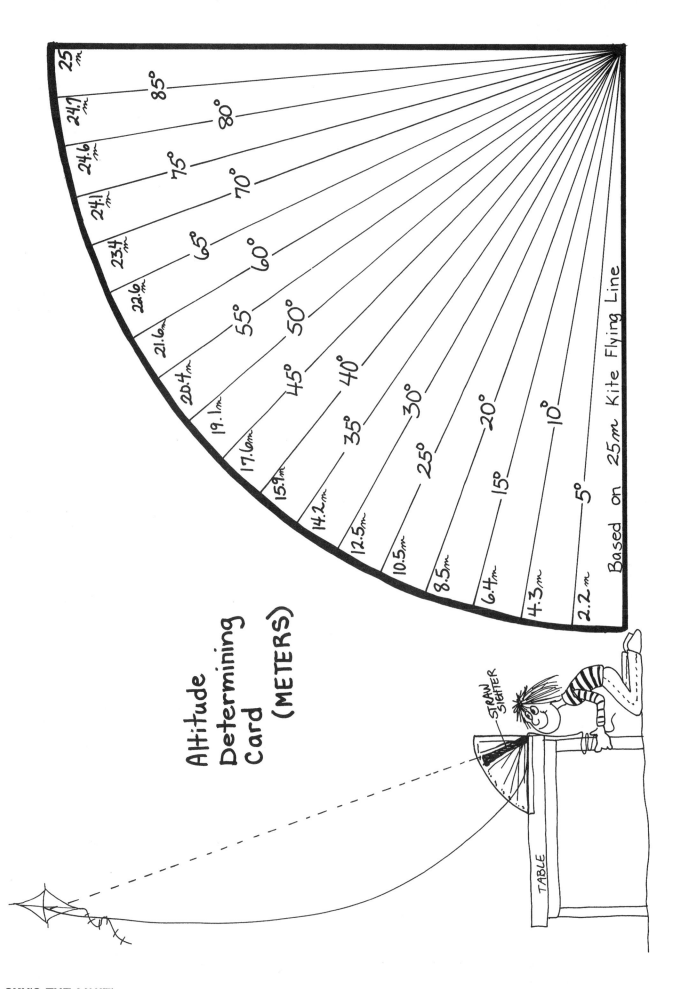

Altitude
Determining
Card
(METERS)

Based on 25m Kite Flying Line

85°
80°
75°
70°
65°
60°
55°
50°
45°
40°
35°
30°
25°
20°
15°
10°
5°
0°

25 m
24.7 m
24.6 m
24.1 m
23.4 m
22.6 m
21.6 m
20.4 m
19.1 m
17.6 m
15.9 m
14.2 m
12.5 m
10.5 m
8.5 m
6.4 m
4.3 m
2.2 m

STRAW SIGHTER

TABLE

CHART B

Ranged Distance to Kite, or Length of Line (Feet)

ALTITUDE of KITE (FEET) / ANGLE (DEGREES) TO KITE

ANGLE	100	150	200	250	300	350	400	450	500	750	1000
5°	8	13	17	22	26	30	34	39	43	64	86
10°	17	26	34	43	52	60	68	78	87	128	174
15°	25	39	50	64	78	90	100	116	129	193	250
20°	35	51	70	85	103	120	140	153	170	256	340
25°	42	63	84	105	126	147	168	190	211	316	422
30°	50	75	100	125	150	175	200	225	250	375	500
35°	57	84	114	142	169	199	228	256	286	428	570
40°	64	95	128	159	190	223	256	285	320	480	640
45°	70	106	140	176	212	247	280	318	350	528	707
50°	77	115	145	191	230	268	290	345	383	574	766
55°	81	123	162	204	246	286	324	378	406	612	810
60°	86	129	172	216	259	303	348	389	430	648	860
65°	90	136	180	226	273	317	360	408	451	678	906
70°	93	140	187	234	279	328	374	422	470	704	940
75°	96	144	192	241	288	338	385	434	482	724	965
80°	98	147	196	246	295	344	393	442	492	738	980
85°	99	149	198	247	297	347	396	446	495	744	990
90°	100	150	200	250	300	350	400	450	500	750	1000

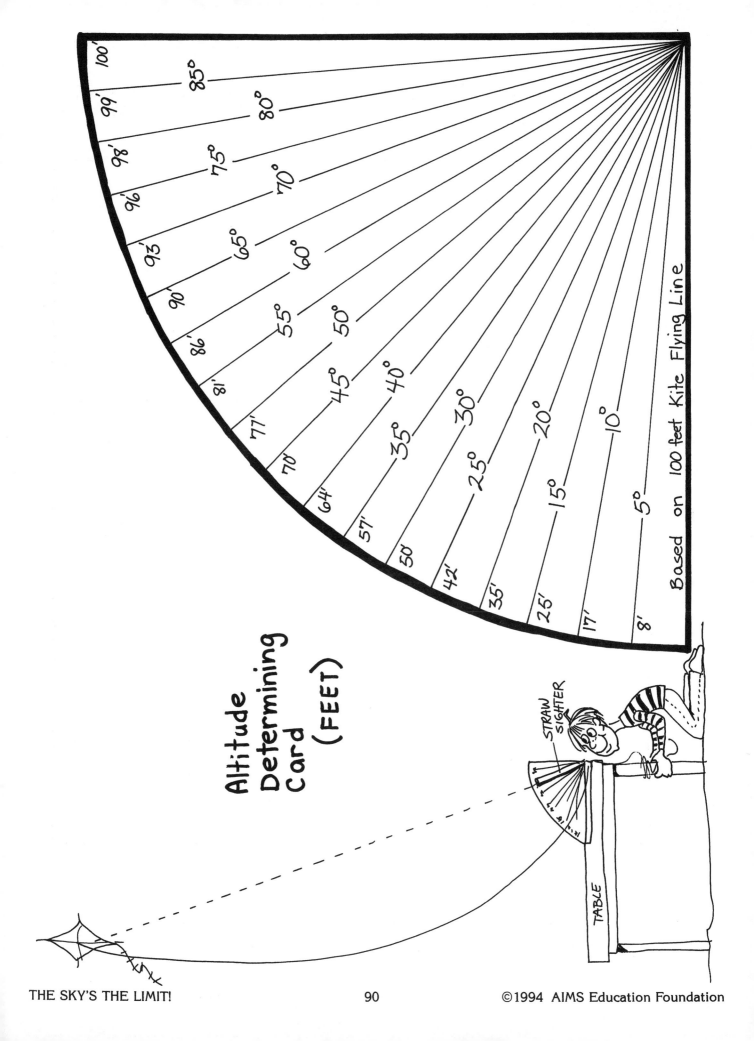

Altitude Determining Card (FEET)

100' 85°
99'   80°
98'   75°
96'   70°
93'   65°
90'   60°
86'   55°
81'   50°
77'   45°
70'   40°
64'   35°
57'   30°
50'   25°
42'   20°
35'   15°
25'   10°
17'   5°
8'

Based on 100 feet Kite Flying Line

STRAW SIGHTER

TABLE

Go Fly A Kite Recording Chart

Altitude of Kite

LENGTH OF KITE STRING USED IN FLYING

KITE WEIGHT _____

DATE OF FLIGHT _____

WIND DIRECTION _____

WIND SPEED _____

A Paperfold Kite

Here is a folded-paper kite that will lift your spirits. It can be made and airborne in a couple of minutes. All that is needed is a sheet of paper, a stapler, scissors, a small piece of tape, a hole punch, and a spool of sewing thread.

This design has no sticks, fins, rudders, tails, stabilizers, or other additions. It is not violently precise, for the kite can made from either a square or oblong piece of paper, from about 8" x 8" and on up. The proportion of width to length should not exceed four to three, but square is optimum.

The kite requires one center fold. If your paper is rectangular, begin folding so that the short sides meet. Take care to crease the paper from the top to a point a little more than one third of the way down. Do <u>NOT</u> fold it all the way from top to bottom.

Carefully bend (don't crease) the top corners down so that they meet at the fold (the keel of the glider) about 4 cm down from the top. The points of the corners should be about 1 cm below the keel. Carefully staple the two corners to the keel with a single staple placed parallel to the keel just above its bottom edge.

Next, mark a spot near the bottom part of the center fold. Reinforce this area with a piece of tape. Punch a hole at the mark. Tie the thread to the kite at this point. You are now ready to fly your kite.

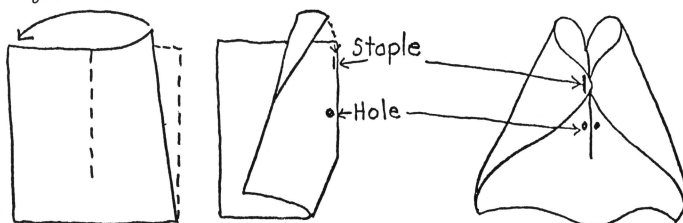

Staple

Hole

DUTCH KITE

The Dutch Kite incorporates the basic elements needed for flight — a spine, a crosspiece, a bridle to hold the kite face to the wind, and a tail for stability. It flies easily in light breezes or heavy winds. It is a tolerant kite and even though a less than perfect model, will still fly.

Materials needed:
- Paper (8½" x 11")
- Glue
- Paper hole punch
- Crepe paper for tail (1" x 4')
- Spool of sewing thread

Making the Kite

1. Fold and crease the paper as illustrated. Fold and crease again, 3/4 inches from center fold. This forms the spine of the kite. Open so the folded section forms a ridge up the middle. This will be the front of the kite, the side you see when you are flying it.

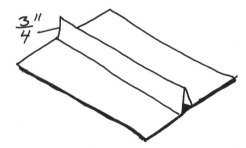

2. Flatten the paper. Make a fold three inches from the top (see illustration). Fold and crease 3/4 inches from this fold to make the cross-folded section. Unfold and cut out the area as illustrated. Save the cutout piece. Flatten the paper, back side up.

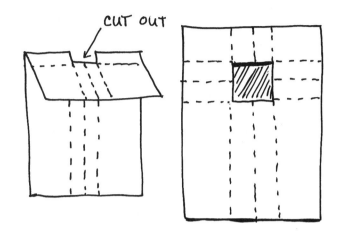

CUT OUT

3. Put glue on the top and bottom sections of the long middle fold (between the fold lines), but not on the cross-folded section. Fold in center, spreading sides out flat, thus gluing the spine of the kite. Now glue the inside of the two cross sections.

4. Glue the cutout piece of paper from *Step 2* so that it joins together the two top sections.

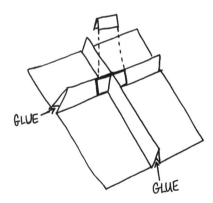

GLUE

GLUE

5. Take a scrap piece of paper (2" x 2½") and fold in the middle in the direction illustrated.

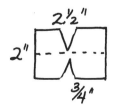

2½"

2"

3/4"

Cut a 3/4" notch through the center of the paper. Glue this to the long center spine to connect the spines.

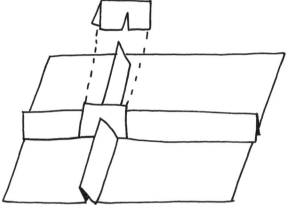

6. Punch two holes for the bridle.

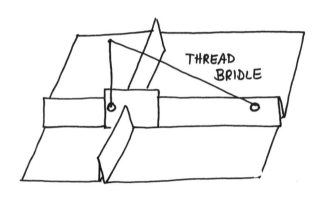

THREAD
BRIDLE

Attach kite flying string.

7. Glue the 1" x 4' tail of crepe paper to the back of the kite.

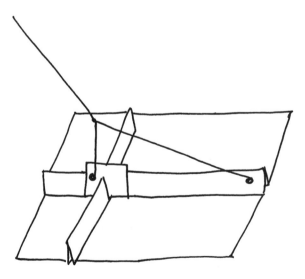

YOUR KITE IS READY TO FLY!

3-STRAW KITE

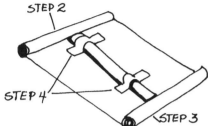

STEP 2

STEP 4

STEP 3

BRIDLE

TAILS

SLIP KNOT

3-Straw Kite

Often we forget that kites need not be complex or expensive. Here is a kite design that is both simple and efficient. Best of all, the cost is quite low!

Materials needed:
Two sheets of paper
Three plastic straws
Glue
Tape
Kite string or thread
Hole punch

Making the Kite

Step 1: Cut one sheet of paper to size as follows: length = length of straw + six cm; width = length of straw.
Step 2: Fold and glue the top edge of paper over a straw.
Step 3: Fold and glue bottom edge over a straw. Leave one straw length between top and bottom
Step 4: Tape the third straw vertically in the center of the back side of kite.
Step 5: Mark the holes for the bridle at the positions shown. Reinforce these marked areas with tape and punch the holes. Tie the bridle string so that it is about 10 cm to 15 cm from the face of kite.
Step 6: Attach kite string to bridle as shown.
Step 7: To make kite tails, cut 2.5 cm wide strips from the second sheet of paper, glue to kite. In light breezes, use fewer tail sections. In strong breezes, use more tail strips for added stability.

FLIGHT DATA SHEET

Project Manager _____

Sponsoring Firm _____

Address _____

_____

Testing and Evaluation Form 192021 A

PREFLIGHT INFORMATION

_____

_____

Have students select one of the kite models they have made for use in this section. Encourage them to choose different versions.

MODE_ _CATIONS

Length _____

Width _____

Shape _____

Area _____

Weight _____

Signature

Encourage students to measure precisely, using metric units, eg. 25.6 cm

Materials _____

Bonding Agent _____

___ent of Flying String _____

___ No _____ Specify _____

Design Engineer

CONSTRUCTION PROCESS

_____

_____

_____

_____

_____

_____

Signature

Plea__ ___low the construction process, making particular ment__ __ difficulties encountered or special manufacturing techniques.

As a Language Arts activity, have students write a paragraph detailing their experiences in building the chosen kite model.

Construction Engineer

COLOR SCHEME _____

DISTINGUISHING MARKS, IF ANY _____

BEAUFORT NUMBER
AND LAND WIND EFFECT

(For use with section 22W)

m.p.h.	Beaufort Number	Land—wind effect	Designation of wind
0–1	0	Calm; smoke rises vertically	Light
1–3	1	Wind direction just shown by smoke	Light
4–7	2	Wind felt on face; leaves rustle	Light
8–12	3	Leaves and small twigs in motion	Gentle
13–18	4	Dust rises; small branches move	Moderate
19–24	5	Small trees in leaf begin to sway	Fresh
25–31	6	Large branches move	Strong

PREFLIGHT CHECKLIST

(In order to comply with insurance regulations, this section must be completed and initialed by the flight engineer prior to flight.)

- ☐ 1. _____
- ☐ 2. _____
- ☐ 3. _____
- ☐ 4. _____
- ☐ 5. _____
- ☐ 6. _____
- ☐ 7. _____
- ☐ 8. _____
- ☐ 9. _____
- ☐ 10. _____

Initialed   Flight Engineer

As a reinforcement to follow class discussion, have students write in the safety rules for flying kites. Check them off on-site when actually flying.

Section Number 22W	WEATHER CONDITIONS			
	Flight No. 1	Flight No. 2	Flight No. 3	Flight No. 4
Wind Direction				
*Wind Velocity/Beaufort No.				
*Wind Velocity/Range or Actual				
Humidity				
Temperature				
Visibility				
Remarks				
Date and Verification †				

*USE TABLE PAGE 1. Look for appropriate description of effect; find Beaufort Number; record range (eg. No. 5 indicates 19—24 mph.) or call the Weather Bureau. † Initial—Flight Engineer

Add Written Evaluation On Reverse Side		FLIGHT DATA		Section 22FDAK	
		Flight No. 1, / /	Flight No. 2, / /	Flight No. 3, / /	Flight No. 4, / /
A L T I T U D E	Estimated Altitude				
	Angle				
	Computed Altitude				
FLIGHT DURATION					
STABILITY INFORMATION					
FLIGHT PERFORMANCE					

You may decide to select a specific day for completing this section or have students use previously collected information.

Initial

THE SKY'S THE LIMIT!   97   ©1994 AIMS Education Foundation

Official Use Only	AW	WP
	LE	ZZ

FLIGHT DATA SHEET

Project Manager _____

Sponsoring Firm _____

Address _____

_____

Testing and Evaluation Form 192021 A

PREFLIGHT INFORMATION

Kite Name _____

Kite Number _____

Type of Kite _____

Date Completed _____

MODEL SPECIFICATIONS _____

Length _____

Width _____

Shape _____

Area _____

Weight _____

Signature

Materials _____

_____

Bonding Agent _____

Attachment of Flying String _____

Tail   Yes _____ No _____ Specify _____

Design Engineer

CONSTRUCTION PROCESS

Please describe below the construction process, making particular mention of difficulties encountered or special manufacturing techniques.

_____

_____

_____

_____

_____

_____

Signature   Construction Engineer

COLOR SCHEME _____

DISTINGUISHING MARKS, IF ANY _____

BEAUFORT NUMBER
AND LAND WIND EFFECT

(For use with section 22W)

m.p.h.	Beaufort Number	Land—wind effect	Designation of wind
0–1	0	Calm; smoke rises vertically	Light
1–3	1	Wind direction just shown by smoke	Light
4–7	2	Wind felt on face; leaves rustle	Light
8–12	3	Leaves and small twigs in motion	Gentle
13–18	4	Dust rises; small branches move	Moderate
19–24	5	Small trees in leaf begin to sway	Fresh
25–31	6	Large branches move	Strong

PREFLIGHT CHECKLIST

(In order to comply with insurance regulations, this section must be completed and initialed by the flight engineer prior to flight.)

☐ 1._____
☐ 2._____
☐ 3._____
☐ 4._____
☐ 5._____
☐ 6._____
☐ 7._____
☐ 8._____
☐ 9._____
☐ 10._____

Initialed   Flight Engineer

WEATHER CONDITIONS

Section Number 22W

	Flight No. 1	Flight No. 2	Flight No. 3	Flight No. 4
Wind Direction				
*Wind Velocity/Beaufort No.				
*Wind Velocity/Range or Actual				
Humidity				
Temperature				
Visibility				
Remarks				
Date and Verification †				

*USE TABLE PAGE 1. Look for appropriate description of effect; find Beaufort Number; record range (eg. No. 5 indicates 19–24 mph.) or call the Weather Bureau. † Initial–Flight Engineer

FLIGHT DATA

Add Written Evaluation On Reverse Side   Section 22FDAK

		Flight No. 1, / /	Flight No. 2, / /	Flight No. 3, / /	Flight No. 4, / /
ALTITUDE	Estimated Altitude				
	Angle				
	Computed Altitude				
FLIGHT DURATION					
STABILITY INFORMATION					
FLIGHT PERFORMANCE					

Initial·

THE SKY'S THE LIMIT!   99   ©1994 AIMS Education Foundation

BERNOULLI WAS A "BIRD-BRAIN!"

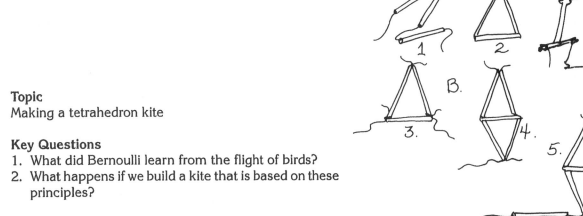

Topic
Making a tetrahedron kite

Key Questions
1. What did Bernoulli learn from the flight of birds?
2. What happens if we build a kite that is based on these principles?

Focus
After building and flying a tetrahedron kite, the students should be able to explain how the kite flies in terms of Bernoulli's Principle and demonstrate the mathematical importance of the tetrahedron.

Math
NCTM Standards
- *Represent and solve problems using geometric models*
- *Extend their understanding of the concepts of perimeter, area, volume, angle measure, capacity, and weight and mass*

Geometry
    plane, solid
Using formulas
    area
    volume
Ratios
Measuring
    length

Science
Project 2061 Benchmarks
- *Mathematics provides a precise language for science and technology–to describe objects and events, to characterize relationships between variables and to argue logically.*
- *Many objects can be described in terms of simple plane figures and solids. Shapes can be compared in terms of concepts such as parallel and perpendicular, congruence and similarity, and symmetry. Symmetry can be found by reflection, turns, or slides.*

Physical science
    Bernoulli's Principle

Processes
Observing
Predicting
Collecting and recording data
Generalizing

Materials
For each kite:
    string
    24 drinking straws
    1 sheet tissue paper
    glue
    tape
    ruler
    protractor

Background Information
We live at the bottom of an ocean–an ocean of air. This air presses on everything. We call this *air pressure*. Bernoulli discovered that air doesn't press as hard when it is moving. The faster it moves, the less it presses. Because an airplane wing is curved along the top and flat on the bottom, the air has to go faster over the top, decreasing the air pressure on the top of the wing. Since this moving air does not press down as hard as the air underneath it, *lift* is created, helping the wing, and the plane to stay up.

Now, Bernoulli had to discover this principle of aerodynamics himself. Birds have always had it built it!! They are natural models of Bernoulli's Principle. Furthermore, the bird's wings are used to "channel" the air moving over its curved body, making the air move that much faster because it has less room in which to move. The kite we are going to make imitates this design of nature.

The kite is built of tetrahedra–solid geometric figures constructed of four equilateral triangles. Since four of them are put together to make the basic kite, it is itself shaped like a tetrahedron. There are some advantages to this design:

- the tetrahedron is one of the most stable of all solid shapes
- by covering only two sides of each tetrahedron, we can channel the air just as a bird does with its wings. In flying their kites, your students may be able to suggest other advantages.

Management

1. This activity works best with students working in pairs.
2. Demonstrate the construction of the first tetrahedron as students follow your lead.
3. The kites can be constructed in two stages of approximately one hour each. On the first day, construct the four tetrahedra. On the second day, cut the tissue paper gores, cover the tetrahedra, and assemble.
4. If using white tissue paper, consider allowing extra time for decorating the gores before covering the tetrahedra frames.
5. Helpful hints for constructing:
   a. Pre-cut the strings
   b. Use square knots. Joints can be taped for added strength.
   c. You can use extra, smaller straws inside the frame straws to prevent bending, but the added weight may be detrimental.
   d. Repairs to bent straws can be made by slitting an extra straw, slipping it over the bent straw, and taping in place.
   e. If allowable, use rubber cement instead of glue as the application is easy and won't cause the colored tissue paper to bleed.
6. Have students provide their own string for flying. Drinking straws may be available from the cafeteria. The gore pattern may have to be adjusted for smaller straws.

Procedure

1. Construct the tetrahedron frames as directed. Once students know the construction techniques, have them record their measurements, etc. on the student activity sheet.
2. Cut the gores from tissue paper.
3. Put the gores on the tetrahedron frames.
4. Arrange the covered frames into a larger tetrahedron and tie together as directed.
5. Attach the kite string and go fly kites.

Discussion

1. What is the ratio between the number of covered and uncovered sides?

2. What is the ratio between the covered and uncovered surface area?
3. What is the ratio between the volume of one of the four small tetrahedra to the whole kite?
4. If you wanted to enlarge the kite and maintain a similar shape, how many more tetrahedra would you have to add? Explain your answer.
5. How do you know the orientation of the kite for flying? [The tetrahedra should form a V to channel the air like the wings of a bird.]

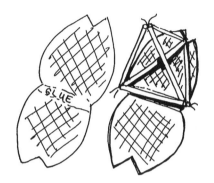

BERNOULLI WAS A "BIRD-BRAIN!"

1. Making the tetrahedron kite

For each tetrahedron frame, you will need: 6 straws, 1 piece of string about 1 meter in length, and 2 pieces of string about 35 cm in length.

a. Thread the meter length of string through three straws and tie them into a triangle.

b. Tie each of the shorter strings to a corner of the triangle. Slip a straw onto each string and tightly tie the two strings together.

c. Slip a third straw onto one of the loose ends of string. Tie the end of the string to the point of the triangle that has nothing yet tied to it. This should complete your first tetrahedron frame.

d. Use the same procedure to make three more tetrahedra.

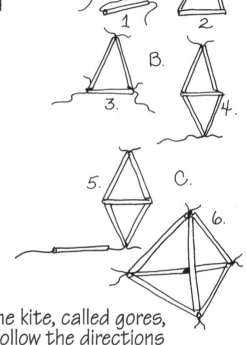

2. Cutting out the gores

The sections which make up the coverings of the kite, called gores, can all be cut from one piece of tissue paper. Follow the directions given on the gore pattern piece.

3. Mounting the gores onto the frames

a. Put a small amount of glue along the fold in the center of a gore. Lay any one straw of a tetrahedron along this glued fold.

b. Cover both of the tissue paper flaps lying outside the straw frame with glue. Fold the flaps over the straws to glue them down inside the triangle.

c. Pivot the tetrahedron frame so that the opposite side of it comes down on the other side of the gore. Glue the flaps as before.

d. Cover the other three tetrahedra in the same way. Notice how they look like birds.

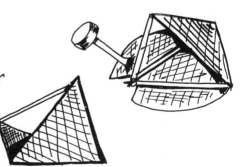

BERNOULLI WAS A "BIRD BRAIN"!

4. Putting the kite together

a. Arrange three of the tetrahedra in a triangle as shown. Be sure that all the gores are facing the same direction.

b. Using three pieces of string, tightly tie the three tetrahedra together at the three places where they touch each other.

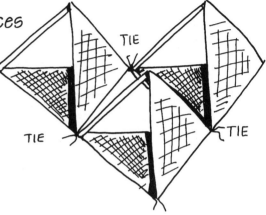

c. Place the fourth tetrahedron above the triangle formed by the first three. Using three more pieces of string, tie the fourth tetrahedron in place.

5. Getting ready to fly

a. Cut off all leftover strings. Tie a string for flying at the "head" of one of the lower "birds," or tetrahedra.

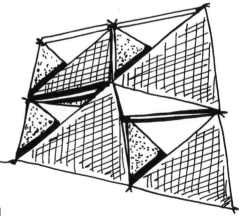

To Think About:
1. What is the ratio between the number of covered and uncovered sides?
2. What is the ratio between the covered and uncovered surface area?
3. What is the ratio between the volume of one of the four small tetrahedra to the whole kite?
4. If you wanted to enlarge the kite and maintain a similar shape, how many more tetrahedra would you have to add?

BERNOULLI WAS A BIRDBRAIN

You are working with triangles and the solids that can be constructed from them—tetrahedra. Working this sheet as you go will help you find out many interesting principles of geometry. You will need a ruler and a protractor.

1. You have just connected 3 straws to make a triangle.
   Record all the facts you know about it:
   Length of a side = _____   All side are _____
   Measurement of an angle = _____   All angles are _____
   Height of the triangle = _____   Sum of the measurements
   Area of the triangle = _____   of the angles _____

   This type of triangle is called an _____ triangle.

2. By adding two more straws, you have a rhombus.
   Record all the facts you know about it.

   A rhombus is a quadrilateral because it has four (quad-)
       sides (-lateral).

3. When you add the last straw, you have made a tetrahedron.
   This is a three-dimensional figure having length, width, and height.

   How many triangular faces does it have? _____   Are they all   ? _____
   What is its volume?_____
   What is the ratio between covered sides and uncovered sides? _____

4. When the kite is assembled, what shape is it? _____
   The entire kite is "similar" to one cell. What is the same about it? _____
   What is different about it? _____
   What is the ratio between covered and uncovered triangles in the
   whole kite? _____
   What is the surface area of the whole kite? _____
   What is the volume of the whole kite? _____
   What is the ratio between the surface area of one cell and the whole kite?

   _____
   What is the ratio between the volume of one cell and the whole kite? _____

TETRAHEDRAL KITE-PATTERN FOR TISSUE PAPER GORE SECTIONS

1. Cut out this pattern, including the broken line above.

2. Fold a full sheet of tissue paper in half, then quarters, and finally eighths.

3. Lay this pattern on the folded tissue paper with the broken line along the fold.

4. Cut the tissue paper with scissors all around this pattern **EXCEPT** along the fold. **DO NOT CUT ALONG THE FOLD!**

5. Unfold the tissue. You should have the four gore sections needed for one kite.

24cm

18cm

BERNOULLI WAS A BIRD BRAIN.

The AIMS Program

AIMS is the acronym for "Activities Integrating Mathematics and Science." Such integration enriches learning and makes it meaningful and holistic. AIMS began as a project of Fresno Pacific University to integrate the study of mathematics and science in grades K-9, but has since expanded to include language arts, social studies, and other disciplines.

AIMS is a continuing program of the non-profit AIMS Education Foundation. It had its inception in a National Science Foundation funded program whose purpose was to explore the effectiveness of integrating mathematics and science. The project directors in cooperation with 80 elementary classroom teachers devoted two years to a thorough field-testing of the results and implications of integration.

The approach met with such positive results that the decision was made to launch a program to create instructional materials incorporating this concept. Despite the fact that thoughtful educators have long recommended an integrative approach, very little appropriate material was available in 1981 when the project began. A series of writing projects have ensued and today the AIMS Education Foundation is committed to continue the creation of new integrated activities on a permanent basis.

The AIMS program is funded through the sale of this developing series of books and proceeds from the Foundation's endowment. All net income from program and products flows into a trust fund administered by the AIMS Education Foundation. Use of these funds is restricted to support of research, development, and publication of new materials. Writers donate all their rights to the Foundation to support its on-going program. No royalties are paid to the writers.

The rationale for integration lies in the fact that science, mathematics, language arts, social studies, etc., are integrally interwoven in the real world from which it follows that they should be similarly treated in the classroom where we are preparing students to live in that world. Teachers who use the AIMS program give enthusiastic endorsement to the effectiveness of this approach.

Science encompasses the art of questioning, investigating, hypothesizing, discovering, and communicating. Mathematics is a language that provides clarity, objectivity, and understanding. The language arts provide us powerful tools of communication. Many of the major contemporary societal issues stem from advancements in science and must be studied in the context of the social sciences. Therefore, it is timely that all of us take seriously a more holistic mode of educating our students. This goal motivates all who are associated with the AIMS Program. We invite you to join us in this effort.

Meaningful integration of knowledge is a major recommendation coming from the nation's professional science and mathematics associations. The American Association for the Advancement of Science in *Science for All Americans* strongly recommends the integration of mathematics, science, and technology. The National Council of Teachers of Mathematics places strong emphasis on applications of mathematics such as are found in science investigations. AIMS is fully aligned with these recommendations.

Extensive field testing of AIMS investigations confirms these beneficial results.

1. Mathematics becomes more meaningful, hence more useful, when it is applied to situations that interest students.
2. The extent to which science is studied and understood is increased, with a significant economy of time, when mathematics and science are integrated.
3. There is improved quality of learning and retention, supporting the thesis that learning which is meaningful and relevant is more effective.
4. Motivation and involvement are increased dramatically as students investigate real-world situations and participate actively in the process.

We invite you to become part of this classroom teacher movement by using an integrated approach to learning and sharing any suggestions you may have. The AIMS Program welcomes you!

AIMS Education Foundation Programs

A Day with AIMS

Intensive one-day workshops are offered to introduce educators to the philosophy and rationale of AIMS. Participants will discuss the methodology of AIMS and the strategies by which AIMS principles may be incorporated into curriculum. Each participant will take part in a variety of hands-on AIMS investigations to gain an understanding of such aspects as the scientific/mathematical content, classroom management, and connections with other curricular areas. *A Day with AIMS* workshops may be offered anywhere in the United States. Necessary supplies and take-home materials are usually included in the enrollment fee.

A Week with AIMS

Throughout the nation, AIMS offers many one-week workshops each year, usually in the summer. Each workshop lasts five days and includes at least 30 hours of AIMS hands-on instruction. Participants are grouped according to the grade level(s) in which they are interested. Instructors are members of the AIMS Instructional Leadership Network. Supplies for the activities and a generous supply of take-home materials are included in the enrollment fee. Sites are selected on the basis of applications submitted by educational organizations. If chosen to host a workshop, the host agency agrees to provide specified facilities and cooperate in the promotion of the workshop. The AIMS Education Foundation supplies workshop materials as well as the travel, housing, and meals for instructors.

AIMS One-Week Perspectives Workshops

Each summer, Fresno Pacific University offers AIMS one-week workshops on its campus in Fresno, California. AIMS Program Directors and highly qualified members of the AIMS National Leadership Network serve as instructors.

The Science Festival and the Festival of Mathematics

Each summer, Fresno Pacific University offers a Science Festival and a Festival of Mathematics. These festivals have gained national recognition as inspiring and challenging experiences, giving unique opportunities to experience hands-on mathematics and science in topical and grade-level groups. Guest faculty includes some of the nation's most highly regarded mathematics and science educators. Supplies and take-home materials are included in the enrollment fee.

The AIMS Instructional Leadership Program

This is an AIMS staff-development program seeking to prepare facilitators for leadership roles in science/math education in their home districts or regions. Upon successful completion of the program, trained facilitators may become members of the AIMS Instructional Leadership Network, qualified to conduct AIMS workshops, teach AIMS in-service courses for college credit, and serve as AIMS consultants. Intensive training is provided in mathematics, science, process and thinking skills, workshop management, and other relevant topics.

College Credit and Grants

Those who participate in workshops may often qualify for college credit. If the workshop takes place on the campus of Fresno Pacific University, that institution may grant appropriate credit. If the workshop takes place off-campus, arrangements can sometimes be made for credit to be granted by another institution. In addition, the applicant's home school district is often willing to grant in-service or professional-development credit. Many educators who participate in AIMS workshops are recipients of various types of educational grants, either local or national. Nationally known foundations and funding agencies have long recognized the value of AIMS mathematics and science workshops to educators. The AIMS Education Foundation encourages educators interested in attending or hosting workshops to explore the possibilities suggested above. Although the Foundation strongly supports such interest, it reminds applicants that they have the primary responsibility for fulfilling *current* requirements.

For current information regarding the programs described above, please complete the following:

Information Request

Please send current information on the items checked:

___ *Basic Information Packet* on AIMS materials
___ *Festival of Mathematics*
___ *Science Festival*
___ *AIMS Instructional Leadership Program*

___ *AIMS One-Week Perspectives* workshops
___ *A Week with AIMS* workshops
___ Hosting information for *A Day with AIMS* workshops
___ Hosting information for *A Week with AIMS* workshops

Name _____ Phone _____

Address _____

       Street   City   State   Zip

AIMS Program Publications

GRADES K-4 SERIES

Bats Incredible!
Brinca de Alegria Hacia la Primavera con las Matemáticas y Ciencias
Cáete de Gusto Hacia el Otoño con la Matemáticas y Ciencias
Cycles of Knowing and Growing
Fall Into Math and Science
Field Detectives
Glide Into Winter With Math and Science
Hardhatting in a Geo-World (Revised Edition, 1996)
Jaw Breakers and Heart Thumpers (Revised Edition, 1995)
Los Cincos Sentidos
Overhead and Underfoot (Revised Edition, 1994)
Patine al Invierno con Matemáticas y Ciencias
Popping With Power (Revised Edition, 1996)
Primariamente Física (Revised Edition, 1994)
Primarily Earth
Primariamente Plantas
Primarily Physics (Revised Edition, 1994)
Primarily Plants
Sense-able Science
Spring Into Math and Science
Under Construction

GRADES K-6 SERIES

Budding Botanist
Critters
El Botanista Principiante
Exploring Environments
Mostly Magnets
Ositos Nada Más
Primarily Bears
Principalmente Imanes
Water Precious Water

GRADES 5-9 SERIES

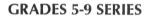

Actions with Fractions
Brick Layers
Conexiones Eléctricas
Down to Earth
Electrical Connections
Finding Your Bearings (Revised Edition, 1996)
Floaters and Sinkers (Revised Edition, 1995)
From Head to Toe
Fun With Foods
Gravity Rules!
Historical Connections in Mathematics, Volume I
Historical Connections in Mathematics, Volume II
Historical Connections in Mathematics, Volume III
Just for the Fun of It!
Machine Shop
Magnificent Microworld Adventures
Math + Science, A Solution
Off the Wall Science: A Poster Series Revisited
Our Wonderful World
Out of This World (Revised Edition, 1994)
Pieces and Patterns, A Patchwork in Math and Science
Piezas y Diseños, un Mosaic de Matemáticas y Ciencias
Proportional Reasoning
Soap Films and Bubbles
Spatial Visualization
The Sky's the Limit (Revised Edition, 1994)
The Amazing Circle, Volume 1
Through the Eyes of the Explorers:
    Minds-on Math & Mapping
What's Next, Volume 1
What's Next, Volume 2
What's Next, Volume 3

For further information write to:

AIMS Education Foundation • P.O. Box 8120 • Fresno, California 93747-8120
www.AIMSedu.org/ • Fax 559•255•6396

We invite you to subscribe to *AIMS!*

Each issue of *AIMS* contains a variety of material useful to educators at all grade levels. Feature articles of lasting value deal with topics such as mathematical or science concepts, curriculum, assessment, the teaching of process skills, and historical background. Several of the latest AIMS math/science investigations are always included, along with their reproducible activity sheets. As needs direct and space allows, various issues contain news of current developments, such as workshop schedules, activities of the AIMS Instructional Leadership Network, and announcements of upcoming publications.

AIMS is published monthly, August through May. Subscriptions are on an annual basis only. A subscription entered at any time will begin with the next issue, but will also include the previous issues of that volume. Readers have preferred this arrangement because articles and activities within an annual volume are often interrelated.

Please note that an *AIMS* subscription automatically includes duplication rights for one school site for all issues included in the subscription. Many schools build cost-effective library resources with their subscriptions.

YES! I am interested in subscribing to AIMS

Name _____   Home Phone _____

Address _____   City, State, Zip _____

Please send the following volumes (subject to availability):

_____ Volume V	(1990-91)	$30.00	_____ Volume X	(1995-96)	$30.00
_____ Volume VI	(1991-92)	$30.00	_____ Volume XI	(1996-97)	$30.00
_____ Volume VII	(1992-93)	$30.00	_____ Volume XII	(1997-98)	$30.00
_____ Volume VIII	(1993-94)	$30.00	_____ Volume XIII	(1998-99)	$30.00
_____ Volume IX	(1994-95)	$30.00	_____ Volume XIV	(1999-00)	$30.00

_____ **Limited offer: Volumes XIV & XV (1999-2001) $55.00**
(Note: Prices may change without notice)

Check your method of payment:

❏ Check enclosed in the amount of $ _____

❏ Purchase order attached (Please include the P.O.#, the authorizing signature, and position of the authorizing person.)

❏ Credit Card   ❏ Visa   ❏ MasterCard   Amount $ _____

Card # _____   Expiration Date _____

Signature _____   Today's Date _____

Make checks payable to **AIMS Education Foundation.**
Mail to *AIMS* Magazine, P.O. Box 8120, Fresno, CA 93747-8120.
Phone (559) 255-4094 or (888) 733-2467 FAX (559) 255-6396
AIMS Homepage: http://www.AIMSedu.org/

AIMS Duplication Rights Program

AIMS has received many requests from school districts for the purchase of unlimited duplication rights to AIMS materials. In response, the AIMS Education Foundation has formulated the program outlined below. There is a built-in flexibility which, we trust, will provide for those who use AIMS materials extensively to purchase such rights for either individual activities or entire books.

It is the goal of the AIMS Education Foundation to make its materials and programs available at reasonable cost. All income from the sale of publications and duplication rights is used to support AIMS programs; hence, strict adherence to regulations governing duplication is essential. Duplication of AIMS materials beyond limits set by copyright laws and those specified below is strictly forbidden.

Limited Duplication Rights

Any purchaser of an AIMS book may make up to *200 copies* of any activity in that book for use at *one school site*. Beyond that, rights must be purchased according to the appropriate category.

Unlimited Duplication Rights for Single Activities

An individual or school may purchase the right to make an unlimited number of copies of a single activity. The royalty is $5.00 per activity per school site.

Examples: 3 activities x 1 site x $5.00 = $15.00
9 activities x 3 sites x $5.00 = $135.00

Unlimited Duplication Rights for Entire Books

A school or district may purchase the right to make an unlimited number of copies of a single, *specified* book. The royalty is $20.00 per book per school site. This is in addition to the cost of the book.

Examples: 5 books x 1 site x $20.00 = $100.00
12 books x 10 sites x $20.00 = $2400.00

Magazine/Newsletter Duplication Rights

Those who purchase AIMS (magazine)/*Newsletter* are hereby granted permission to make up to 200 copies of any portion of it, provided these copies will be used for educational purposes.

Workshop Instructors' Duplication Rights

Workshop instructors may distribute to registered workshop participants a maximum of 100 copies of any article and/or 100 copies of no more than eight activities, provided these six conditions are met:

1. Since all AIMS activities are based upon the *AIMS Model of Mathematics* and the *AIMS Model of Learning,* leaders must include in their presentations an explanation of these two models.
2. Workshop instructors must relate the AIMS activities presented to these basic explanations of the AIMS philosophy of education.
3. The copyright notice must appear on all materials distributed.
4. Instructors must provide information enabling participants to order books and magazines from the Foundation.
5. Instructors must inform participants of their limited duplication rights as outlined below.
6. Only student pages may be duplicated.

Written permission must be obtained for duplication beyond the limits listed above. Additional royalty payments may be required.

Workshop Participants' Rights

Those enrolled in workshops in which AIMS student activity sheets are distributed may duplicate a maximum of 35 copies or enough to use the lessons one time with one class, whichever is less. Beyond that, rights must be purchased according to the appropriate category.

Application for Duplication Rights

The purchasing agency or individual must clearly specify the following:
1. Name, address, and telephone number
2. Titles of the books for Unlimited Duplication Rights contracts
3. Titles of activities for Unlimited Duplication Rights contracts
4. Names and addresses of school sites for which duplication rights are being purchased.

NOTE: *Books to be duplicated must be purchased separately and are not included in the contract for Unlimited Duplication Rights.*

The requested duplication rights are automatically authorized when proper payment is received, although a *Certificate of Duplication Rights* will be issued when the application is processed.

Address all correspondence to: **Contract Division**
AIMS Education Foundation
P.O. Box 8120
Fresno, CA 93747-8120

www.AIMSedu.org/
Fax 559 • 255 • 6396